A WESTERN HORS

COW-HORSE CONFIDENCE

A Time-Honored Approach to Stockmanship

By Martin Black with Cynthia McFarland

Edited by
Fran Devereux Smith and Cathy Martindale

Photography by Darrell Dodds

Family Photographs Courtesy Martin Black

Alvord Ranch Branding Photography by Kim Stone

Based on the Original *Western Horseman* Magazine Series
Written and Developed by Jennifer Denison

Cow-Horse Confidence

Published by
WESTERN HORSEMAN® magazine
2112 Montgomery St.
Fort Worth, TX 76107
817-737-6397

www.westernhorseman.com

Design, Typography, and Production
Sandy Cochran Graphic Design
Fort Collins, Colorado

Front and Back Cover Photos by
Darrell Dodds

Printing
Versa Press, Inc.
East Peoria, IL

Manufactured in the United States of America

Fourth Printing: March 2016

ISBN 978-0-7627-7012-0

Dedication and Acknowledgments

I would like to dedicate this to all the good hands who inspired me from the time I was a small boy, but to one in particular, who stood out more to me than any other. Later in my life a good friend asked me who was the best all-around cowboy I ever worked with. Without having to give it much thought, I said, "Ray Hunt."

Although I've traveled a lot and worked horses and cattle around many good hands, Ray is still an inspiration to me. We all know him as a famous horseman, but when I got to know him in my early teenage years, he impressed me as a good stockman, roper, bronc rider and horseshoer, and, among other things, he had an appreciation for fine gear. He could use a reata as well as anyone, he always had a nice bridle, and his horses were always well-groomed. Who looking for a top hand wouldn't be inspired by Ray Hunt?

I also owe gratitude to the people of *Western Horseman*, Darrell Dodds, Cynthia McFarland, my wife, Jennifer, Kim Stone, Emily Kitching and my sister, Sandy Black.

Contents

FOREWORD

I was quite flattered when Martin asked me to write the foreword to this book, but after thinking it over, I realize I might be the only horseman still alive who knew him as a boy.

My first impression of Martin, as he was growing up, was that he had a hell of an ego. Then I realized it was his compelling desire to be the best he could possibly be in the livestock world. As I watched him develop, I knew he could become one of the great vaquero horsemen of his generation, and he is proving me right.

The thing that impresses me the most about Martin is his ability to teach what he knows, and that he is motivated by a desire to preserve this ancient way of horse-handling. He does not want it to be a lost art. Long ago, it was proven to me that horses lucky enough to be handled this way usually turn out well.

I hope the people who read Martin's book will want to use this system to make handier, happier, longer-lasting horses.

Tom Marvel
Nevada
Spring 2010

Tom and Buttons, a mustang, won the open bridle class at the Elko County Fair, circa 1958.

INTRODUCTION

"The West is dead, my friend, but writers hold the seed and what they saw will live and grow again to those who read." C.M. Russell, 1917

In 1882, Kid Russell drifted into Montana, where he worked around the cowboys in the 1880s and 1890s, and later became an artist. The country forever changed after the open range started closing up with homesteaders' fences, and the buffalo and Indians quit roaming free. Russell then wrote to a friend saying, "It will never be like it used to be."

After hearing stories of my ancestors in Idaho a generation prior to Russell's experiences, I realize each generation can make the same claim. My children and their children likely will say the same. That's part of evolution; we must change with the times, learn from our past and look to the future.

The grass might seem greener on the other side of the fence until we get there; then we find out it's not that much better. We can learn from all the hands we work around, good and bad. It's obvious what we can learn from the good ones, but the bad can point out what we don't need to do. Likewise, we can learn from all our experiences, good and bad.

I would like to share my experiences with you so maybe everyone doesn't have to make as many mistakes as I did. Whatever knowledge I might have didn't come from books and schools. Some of the most important things I have learned came to me while lying in a bedroll after an impressive experience. It might have been a good one or it might have been a bad one, but nonetheless, such experiences have taught me the most. In this book I mention some of the hands that I got to work around who helped me find some of the good experiences.

Hopefully, studying this book can keep you and your horses from having so many bad experiences. It's good to educate yourself with all the information you can.

But without experience, we can't have the true knowledge we need to communicate with livestock. They know only what they learn from experience. They communicate through feel, a feel we can learn only from them through experience; there is no other way to learn that feel for stock.

As Tom Dorrance said, "Nobody knows more about being a horse than a horse." There are so many parallels between horses and cattle; understanding one helps us to understand the other better. We can learn a lot about the horse from a cow, and we can learn a lot about the cow from a horse. At the same time, we can use our horses to work cattle, and we can use cattle to work our horses.

It's so much more interesting for the horses if they see we have a purpose for them. They like to have jobs. If we can learn to use the cows as tools to give the horses purpose, they can have better experiences. There are so many things we can get done with our horses while setting up "a job" for them and letting a cow school them. Their languages are closer to each other's than to ours. People hide a lot of their feelings, but with horses and cattle, that emotion is right out there for anyone to feel. So if we can learn to develop our feel for horses and cattle, and suppress our authoritative demeanor around them, they are more open than ever, and we can all learn more from each other.

My hope in writing this book is to help people find some of the fulfillment they can get from learning a new language to communicate in a way that is fitting for the horse, and also to maintain some things from a tradition that lived in a time before the mechanized revolution. My belief is that we need to hold onto some of the principles that have worked for the horse in the past, and not be in such a hurry to find a faster, easier way.

The past can definitely be improved and it has been, but sometimes it's too easy to get caught up by the momentum of this fast-paced world. Livestock handles at the same speed they did a generation ago, even a thousand generations ago.

God wired us as hunters and made cattle and horses the hunted. They are fully aware of this. On the other hand, man tends to overlook the dynamics of this much too often; doing that becomes an obstacle for our livestock and us. The old ranches, where cowboys lived with the stock, had a different perception than what we see today with recreational riders. There was an opportunity to gain firsthand information, but too often today, information is passed along a gossip line by people far removed from the origin of the message, in this case, the horse.

In my grandfathers' time and their fathers' time before that, horses were their business; my family raised and sold thousands of horses. Today, horsemanship is a big business, and too often the salesmanship and the showmanship supersedes the horsemanship. We need to sort through that and let the horse confirm what works.

I feel very fortunate to have grown up and worked in a time when horses were used as work animals. I learned not only from my family, whose living came from horses, but from other horseman like Bill Van Norman, Ray Hunt, Melvin Jones, Tom Marvel, Tom Dorrance, Gene Lewis and many more not as well known. To these great horsemen I am forever grateful for them sharing with me something that can be so inspiring, fulfilling, frustrating, and discouraging—and all experienced on the same ride.

"I still think about the various horsemen who influenced me and about the reasons why things should be done in certain ways."

1

Life-Shaping Events

From my earliest days, horses and cattle have been a major part of my life. Living a ranching lifestyle and caring for livestock shaped and influenced both my childhood and, eventually, my career. I have the good Lord and my family to thank for that.

On both sides, my family goes back seven generations ranching in Owyhee County, Idaho. The Black family was among the early settlers of the Bruneau Valley, and the Joyces, my mother's family, settled on Sinker Creek in 1862. My mother's family actually has the oldest continuously owned family ranch in Idaho. Their original MJ brand is still used

The Joyce Ranch on Sinker Creek in Owyhee County was established in 1862. My maternal side has the oldest continuously owned family ranch in Idaho.

My great-grandfather, Joe R. Black, grew up spending summers where six buckaroo wagons camped in the 1870s and 1880s.

Grandfather Albert Black was a top roper on the ranch and in the competitive arena.

today and has the distinction of being the oldest continuously used brand in the state.

Joe Black, my great-grandfather, was born in 1875, and grew up in the late 1800s working with the California vaqueros who brought the first cattle into that part of Idaho. He was respected as one of the top horsemen and one of the finest ropers in the Idaho-Oregon-Nevada, or ION, region. Joe followed the time-honored traditions of those early California vaqueros and usually roped with a rawhide reata.

My great-grandfather and his brothers raised thousands of horses through the late 1800s and into the early 1900s. These were Thoroughbreds, which they sold to the U.S. Cavalry and to European governments for military use, as well as to ranches across the American West.

My grandfather Albert continued this tradition of raising quality Thoroughbreds. Albert and my uncle Paul also started colts and working cow horses, and were considered fine horsemen in their day. After World War II, the demand for horses lessened and the market diminished. As a result, the focus of my family's ranching moved more toward cattle, and the horses they raised were mostly dedicated to working those cattle. The lessons I learned from my family about handling horses and stock had a lasting impact and helped mold the man I've become.

Asa Black, my father, was a champion roper and bronc rider. Note the silver bridle on the rope horse.

Working Cattle

My cousins, siblings and I were riding and roping at a young age. When I was still a boy, I realized I couldn't always depend on having someone else around to help me get the job done. I had to work with what I had and adapt to what needed to be done. I might not have another rider or a corral in sight, but I still had to catch a cow and handle her. I had to get more done with less assistance. Much of what I learned back then applies to my philosophy today.

My dad started sending me out to cow camp when I was 8 years old. My great-uncle was in cow camp when he was 14 or 16, starting horses and looking after the cattle all by himself. I had my 14-year-old brother, Terry, with me, so I guess my dad figured the two of us could handle whatever came up.

The cow camps were between 60 and 80 miles from the home ranch, and we didn't have any modern conveniences. There was no Nintendo to play, no football games on TV, no indoor plumbing or electricity, but I was used to this way of living. My dad might haul groceries out to us, or else we went to town to get what we needed. I learned to be pretty self-sufficient from a young age.

We ran about 900 head of mother cows and in high-desert country like that, they scattered quite a ways. All our working hours—and those were long days—revolved around the cattle. Our days were spent moving them, monitoring where they needed to go, keeping fresh water and feed in front of them, and packing salt to them. We also had to keep the right proportion of bulls to the cows.

There were virtually no fences back then, so we were continuously pushing cattle out of country where we didn't want them and pushing them into country where we did want them. We'd have to ride into the neighbors' country taking their cattle back. While we were there, we'd pick up any of our own cattle that had wandered off and put them back where they belonged. Then our cattle could be branded with our irons and bred by our bulls.

In the lower open desert country we'd have green grass only in a given area about two months at a time. That grass might be

James H. Black, my great-great-grandfather, is shown here with a young Thoroughbred stallion. James was an early pioneer raising horses in the Bruneau area.

The back of this photo, taken at the Idaho State Fair in 1900, reads, "Grandpa's rope broke. He retied it and finished in second place." Joe R. Black is on the gray horse.

gone by June, depending on the year, so then we moved cattle to the higher altitudes where there were high meadows and springs. We didn't want our cattle in that country early in the year because we wanted to save that area until after the lower country had dried up.

It was the same with water. The springs, small creeks and reservoirs would dry up in the lower country, so we would have only about 10 percent of the water in that area in the late summer that we had in the early summer. In the high country, water diminished somewhat when the snow melted, but the springs maintained better there than in the low country.

We did a lot of branding through the summer, but we didn't necessarily have one big spring branding with all the neighbors. We might have five calves in a group of cows, so when we got the cows where we wanted them, we just built a fire. We always kept running irons on our saddles and we'd brand the calves with those. By end of the summer, we had all our calves branded.

The yearlings also ran with the cows because there wasn't a separate designated area just for them. Later in the summer,

At the Joyce Livestock branding, children always were part of the family operation.

Doug Black, my uncle, was an all-around cowboy rodeoing in the Northwest and favored riding bucking horses on the ranch.

as we'd push cows over to new grass, we pulled out the yearlings and herded them to a different part of the ranch. There were no fences to hold them, so we were continuously bringing cows and bulls out of areas where we were trying to keep the yearlings, and trying to keep the yearlings out of the cows. By the end of the summer, most of our yearlings were ready to ship.

We had some fenced fields to hold cattle when we got ready to ship, but not big enough areas to hold them up for longer than a month. Because of this we were constantly herding cattle through the summer. Back then, everyone—my uncles, the neighbors—all ranched this way.

The generations before me all came up through the Great Depression and World War II; they lived and operated the ranches like hard times could come back any day. At the time, I just thought it was the way everybody did things. This was just the way things were done; I didn't think it was

Cousin Mike Black is shown here as a young buckaroo on the Gamble Ranch in Nevada in 1948.

Milly is shown here with her younger brother, Paul Black, swinging a rope. Good cowboys start young.

unusual. But as I got older and got out of that country, I saw how other ranches operated, and I realized it was different. The way most ranches are run is for the convenience of the people and because of that, they're losing a lot of tradition.

That's what has inspired me to try to preserve some of those old ways. I don't want to just talk about the "good old days," but rather show people they can still handle cattle this way today. These skills are still useful, even in a modern environment.

Brothers Terry and Tony are shown on the horse Asa used picking up bucking horses at rodeos. The boys apparently picked up girls on him, too.

When I was growing up, we knew almost all of our cattle, if not by name, by sight, because we worked and lived with them all the time. That's missing in many ranching situations today, and a lot of people don't know their cattle at all.

Our cattle weren't really wild even though they lived on the range. They might trot along for a while, but when we got in front of them, we could stop them. Once we let them relax and breathe, they would stand, and we could ride into the herd and rope cattle out of it. That's the way cattle used to be worked all the time; we didn't take them to a corral to work them. They learned that once we settled them, they were more or less parked there, and wouldn't leave until someone pushed them. We trained them this way simply by repetition.

For a good example of this "training," take a closer look at the Charles Russell painting *The Herd Quitter*. One cow has left the herd, and the cowboys chase her down and rope her. They made it harder on her out away from the herd so she wouldn't want to leave on her own again. Trust me, cattle learn fast. That cow wouldn't be likely to bolt from the comfort and safety of the group again.

Handling cattle and horses is very similar. The more we understand one, the easier it is to understand the other. Call it stockmanship, horsemanship, cowmanship—

to me it's all the same. These animals avoid discomfort and seek comfort.

Many people don't realize they can train cattle just like they train horses. Cows can be spoiled to the point that they're always trying to get away and go hide, or they can learn to settle and stand until you ask them to move again.

Horsemanship Mentors

By the time I was a freshman in high school, I'd started a number of colts and already had an appreciation for training young horses. I met Ray Hunt when I was 14, and his methods and philosophies unveiled a new layer of horsemanship for me. The way he handled horses made good sense. I appreciated his ideas enough that when I left home after graduating, I headed to California to work for him when I was 18.

When it came to horsemanship, I didn't have just one mentor; I probably had eight or 10. Along with Ray Hunt, I learned from such notable horsemen as Tom Dorrance, Gene Lewis, Tom Marvel and Bill Van Norman. I also learned a great deal from Melvin Jones, a student of my Uncle Paul, who later became a great reined cow horse competitor.

I worked for these men because I wanted to learn what they could offer. I was with Tom Marvel for six months and never got a paycheck, but I had a roof over my head and he fed me. I didn't go to town and spend money, so I really didn't want for anything and, more importantly, I was learning something every day.

When he was still a fairly young man, my Uncle Paul went to Nevada in the 1940s to help manage the TS Ranch. He recruited quite a few people in our home community of Bruneau, Idaho, to move there to work with him. Although there were some good horsemen in Nevada, the owner of this ranch, Rowland Hill, really made an impact in the area because he was one of the first people who actually was trying to train his help. With most employers, it was just "watch and learn" since nobody would really explain how to do things.

Paul liked the way this owner thought, and in turn tried to help his employees learn to be good horsemen. Call it "mentoring" if you will, but it wasn't common in those days. My uncle wasn't a clinician; we hadn't

As a young man, Gene Lewis started horses on the Joyce Ranch.

Uncle Paul Black was respected as a top horseman in northern Nevada and southwestern Idaho.

Ray Hunt was a top hand on the ranch and with a horse.

heard of that back then, but if he could help someone, he would. He wasn't the macho tough cowboy type and he didn't care to compete, but he cared a lot about horses and he cared about people.

I worked for Bill Van Norman one spring and worked for a neighboring ranch two years. While in high school, a friend of mine and I helped Bill start his colts every year. We learned a lot about colts, and later I learned a lot about showing horses. Bill was one of the smoothest showmen I ever saw.

"Handling cattle and horses is very similar; the more we understand one, the easier it is to understand the other."

It was through Bill that I met Ray Hunt and Tom Dorrance.

I was around Ray Hunt off and on from the time I was 14 years old and worked for him the first winter out of high school. I feel very fortunate to have been around Ray when he was doing a lot of ranch work with his horses before he started giving clinics full-time. His horses were handy in the clinics, but they were really broke with all the ranch work. I have never seen anybody's horses work for him like Ray's horses worked for him then.

Gene Lewis grew up on the Joyce ranch and was related on my mother's side of the family; he and my dad were old friends. Gene was living in California when I met him and shortly after high school, I went and stayed with him to ride jumping horses. At the time I was riding saddle broncs, and Gene not only helped me with good sound horsemanship, but also with some good experience that helped me riding broncs.

Many of those men who went to work for my uncle eventually returned to the Bruneau Valley, and those were the ranchers I grew up around. Paul's insights into horsemanship made a difference in their lives, and in mine, as I learned from these people. I had an exposure at a young age that was unlike either my father's generation or my children's generation.

Later, when I was out on my own, I went to work in Nevada because of the good reputation of horsemen in that area. In addition to meeting some of the men my uncle had trained there, I had the privilege of knowing Tom Marvel. Tom never worked for my Uncle Paul, but he owned a neighboring ranch and was a talented horseman.

Melvin Jones, another top horseman, was one of my uncle's "disciples." He started working for Paul at the age of 16 and stayed in the area. Melvin isn't well known today because he never competed on the national level, but in his region he was considered one of the best. Tom's and Melvin's methods weren't identical, but they were compatible, and I always will value both men for all I learned from them.

To this day, I still think about the various horsemen who influenced me and the reasons why things should be done certain ways. I think it's fair to say that, as an adult, I haven't seen anything that matches some of the horsemanship I saw as a kid. I give those men credit for being every bit as good—if not better than—the horsemen we see today. My mentors' horses didn't have arena-trained backgrounds, but those men could brand cattle and rope on their horses today, and then go in a bridle-horse class tomorrow.

Career Moves

Experience is often the best teacher, and I gained plenty of that during my eight years, from 1984 through 1992, in a management position at the Wine Cup Ranch. The Wine Cup, a 1.2 million-acre operation in northeastern Nevada, ran 400 horses and 15,000 head of cattle.

I'd known Tom Dorrance since the mid-1970s, and he used to visit the ranch during the summer to work with the buckaroos and their horses. We started about 40 colts there every year and got horses ready for the Elko County Stockhorse Show at the same time. Between the different buckaroos and me, we maybe had around 15 or more show horses that Tom would help us advance. From Tom, more than anyone else, I learned to work on myself more and not try to train on the horse. My only regret is not seeing this sooner.

I consider myself fortunate to have worked with and learned from some of the best horsemen and buckaroos throughout Idaho, Oregon and Nevada. All of the knowledge and experience I'd gained through the years came into play when I started my own business in 1993. I was starting colts for reined cow horse and cutting horse trainers, as well as young Thoroughbreds in Kentucky and along the East Coast.

People might wonder about the difference between training stock horses and future racehorses, but I use the same principles and methods, no matter the breed. During the first few weeks, it's really all the same. Starting young horses is still one of the fundamental aspects of my business and one of the things I most enjoy.

When I was working hard at a young age, I didn't realize what a tremendous opportunity I had simply because of the lifestyle I was born into. It definitely wasn't all fun, but I learned some lifelong horsemanship and stock-handling lessons. Today, I enjoy sharing those hard-earned lessons with horse owners and students who want to study and improve their skills.

Even as I'm teaching others, I realize there are still lessons to absorb from the horses and cattle I work with every day. These honest animals speak a silent language that is as clear as the spoken word if I only stop to listen and learn.

Tom Dorrance gave me some pointers on a young horse in the mid-1980s at the Gamble Ranch in Nevada.

"The tradition of the bridle horse is important to me because I think it could easily become a lost art, and I'd like to preserve it."

2

Tradition and Theory

We get praise and criticism from the people in our lives, but the horse truly can give us validation. I think this is one compelling reason why I have been drawn to them since I was young.

In the last century, so much has changed about how horses are used in most of the world. This has been both positive and negative for the animal.

Collecting an assortment of bits is not only a hobby, but also allows the opportunity to offer different choices to each horse until the best fit is found.

Prior to World War II, the first three generations of Blacks shipped thousands of horses through these railroad corrals at Mountain Home.

If you look at the history of the horse, he was used for thousands of years more as a war tool than anything else. Then in California in the 17th century, the horse was developed as a stock horse as much as or more than anywhere in the world. The vaqueros there fine-tuned the bridle horse, the spade-bit horse and the rope horse. Of course, in South America they also developed some of this tradition, but I don't know of any place other than California where horses were used exclusively in this way.

The mechanical revolution brought progress in many areas, but it was the beginning of the end of the working cow horse. Instead of trotting 20 miles to the cattle, cowboys loaded their horses in trailers and hauled them wherever they needed to work.

For the most part, horses weren't really needed after World War II the way they had been before. Starting around the 1960s and 1970s, horses became more of a hobby for Americans, and that is still the main use of most horses today.

Many ranches don't even need horses to do the work. The owners might have horses because their kids are involved in some kind of activity or they're trying to promote a breeding program. But the truth is that not many ranches have horses that don't go to a performance arena at some time, whether it is for 4-H, rodeo or showing. The majority of ranch horses at some point are used for performance.

When I was a kid, this wasn't even a consideration; horses were work tools, plain and simple. We definitely had our favorite horses, and we might have somewhat of an emotional bond with those animals. But in the back of our minds, we always knew that someday we might have to shoot one of those horses if something happened to him. Growing up on a working ranch, I learned at a young age that death is a part of life. As a result, we limited how emotionally attached we got to the animals we worked, whether those were our dogs, our horses or our cattle. No matter how much we appreciated them, we saw them as farm animals and didn't look at them as the equivalent of humans.

I have a lot of respect for horses and their abilities. I definitely have more admiration for them than for other livestock, but I still think of my horses as farm animals.

I see a lot of miserable horses that are coddled and pampered to the point they are practically smothered with affection. I think there should be a happy medium. For example, I don't think horses enjoy being in a stall any more than they really enjoy being outside on a cold, windy day with their tails to the wind. I know I wouldn't be very happy in a 12-by-12 cell, with nothing to do but eat and sleep, with nothing to occupy my thoughts, and unable to run, roll or mingle with others. We feel sorry for horses and think they should be in a barn because we would be warmer there. But we can put a barn out in the middle of the desert, and wild horses won't go to it.

My horses have a run-in, but they don't usually go into it unless they're looking for shade. They were raised in open desert country and deal with the natural elements. I don't believe in being inconsiderate to the horse, but at the same time, I don't think we should smother him.

Bridle-Horse Traditions

When I refer to a "bridle horse," I am speaking of the traditional bridle horse trained in the old vaquero ways, not merely a horse ridden in a bridle.

The tradition of the bridle horse is important to me because I think it could easily become a lost art, and I'd like to preserve it. I'd hate to see it vanish because there is so much value in those old traditions. I'm not trying to go back in time; I'm fully aware that the country is fenced and that horses are bred differently today. But I think we can accept these changes without giving up the tradition. I encourage people who really want to learn more about the making of a bridle horse to study with someone skilled in this art.

The horses you currently see in bridle-horse competitions aren't made the way they were when I was growing up. Very few horses you see in the show pen today are totally honest about going out and doing their jobs. Instead of honesty, many are performing more out of fear and intimidation.

They have been trained with devices to inhibit their head positions and their tails have the nerves deadened to hide expressions of resentment.

"We had one camp where there was a fence on only one side of it. We could leave camp in the morning heading south and ride all day without opening a single gate."

When I was young, we didn't start our horses until they were 3 or 4 years old. Basically, we just gentled them at that age. We rode them a few times and then turned them out for another year. We didn't start really using them until they were almost mature, around 5 years old. By then, a horse was able to put in a full day's work and our "training program" was just ranch work. Those horses were ridden in a snaffle bit

Joe and Frank Black, shown in 1903 with the other buckaroos, corralled horses. There always have been jobs for horses on the ranch.

or hackamore and were, for the most part, fully trained before we ever put them in the bridle. By this, I mean putting a horse in a leverage bit. Training a horse in a hackamore required us to be better horsemen than we had to be when training a horse in a snaffle, let alone with a leverage bit.

"You can learn equine anatomy by reading and studying, but you can't learn horsemanship anywhere else but from a horse. The horse is the only one who can tell you if what you're doing is working."

Today, I see people putting leverage bits on 2-year-olds. Too often, it's more about taking shortcuts than about developing a good finished product. I actually shy away from using the term "finished" because, to me, that means I can't do any more, and I don't think a horse ever gets to the point where he can't be improved.

In the past, a person took more responsibility by developing himself and his skills, by using less in the way of a bridle, such as a hackamore. I think traditional ways tend to favor the horse because the rider must strive to become a better horseman in order to make a better horse. More bridle means more responsibility for the horse to learn what a person wants, while less bridle means more responsibility for the person to learn what the horse needs.

For example, a snaffle bit is more severe than a hackamore, while a leverage bit can be more severe than a snaffle. Any time you're looking for a more severe way of controlling or training the horse, you're not developing your skills so much as you're trying to develop the horse's skills. That's not fair to the horse.

We're supposed to be the more intelligent species, yet we humans often emphasize brawn over brains. I like the saying, "When a man runs out of knowledge, he uses his strength." The secret is to gain more knowledge, not find more strength. Our goals as riders should be to get better and better at presenting the message so that it is easier for the horses to understand. If we do this correctly, we don't put as much

Mattie and Paul Black are shown on two spade-bit horses in 1948 at the Gamble Ranch in Nevada. Paul is riding his horse in the two-rein.

responsibility on the animals to interpret the message.

In the old school, if a horse was ridden in the bridle, that was the ultimate. That person already was riding with one hand and neck-reining the horse; the bridle was just to refine the touch of everything the person already had taught the horse.

I asked my uncle one time how he adjusted his curb strap and he said, "Well, I never really worried about adjusting it because it never came tight."

Our horses weren't in the bridle until they were at least 7 or 8 years old, but by the time they were 10 or 12, they were pretty top horses. They weren't considered "old" until they were 25 or 30. When I was a kid, some of my first saddle horses were almost 30 years old and could still put in a pretty good day's work.

We had enough horses that we never had to ride the same one every day. We might ride one horse once or twice a week, and some of the tough older horses might not even get ridden that much. We saved them for when we really had to put in a long, hard day. They were the ones that took us for bronc rides in the morning, but they'd go all day, and we didn't have to "pedal" them on the way home.

We typically rode a string of horses for a season and then turned them out, rotating strings so they got ridden every other season. Our horses had as much time off as they did working. They might have been used hard when they worked, but they had plenty of time off, and that's why they lasted so long. With a lot of these horses, even if we didn't go for a bronc ride every time we saddled up, it sure felt like we might.

Back then, all the spade bit did was put the refining touch on a "made" horse. It's a different story today. You look in some training barns, and you see every kind of bit you can imagine.

In the past, the buckaroo might experiment with different bits to see what weight and taste the horse preferred, but once a buckaroo found a bit that horse wanted, that was his bit from then on. When a buckaroo sold or bought a horse, his bit was just part of the package.

There was a period in the early 1900s when my granddad was buying a lot of cattle. At that time a good bridle horse wasn't worth more than a few steers. When he was looking at cattle, Granddad would

Here's the headgear on one of my bridle horses.

The headstall and curb strap adjustments should be checked for proper fit.

There's never been a shortage of horses.

notice the better horses, and he'd often negotiate and pick out a horse from the ranch that was selling the cattle. That horse's bridle and bit always came with him.

Those cowboys had a lot of time invested in their horses. They weren't doing that for a paycheck, and they weren't doing it because the ranch owner wanted them to win a competition. They didn't get paid any more for making a good bridle horse. What motivated them was the opportunity to show up at the rodear grounds and have the nicest bridle horse. It wasn't about money; it was a matter of personal pride to refine a horse.

It's a lot like the kid today who spends hours polishing and detailing his car. It's really about what he has personally in the vehicle—emotionally, as well as financially—not about what the market says the car is worth.

Changing Times

For nearly 100 years, my family was in the business of selling horses to the Army and to other ranches. This was before the days of using horses for recreation, so there wasn't really a show-horse market. Even with years of training invested, the average horse

Part of maintaining a horse well mentally is to try to maintain him well physically. It's not much trouble to raise a saddle and let some blood and air circulate to the pressure points of the saddle. We should think about this every time we reposition our tight, hot hats.

wasn't worth much money. It was rare back then to have a horse just for a hobby. There just wasn't much of a high-dollar market for a riding horse or even a good workhorse.

When I talk about those horses not being worth much money, I still have to say the majority of them probably were better bridle horses than most I see today. In the old days, horses weren't judged by pedigree; it was about what the horse could do. What we see today all comes down to paper value. It's all about a young horse's breeding and what his family has earned. That's how people put value on a lot of horses now.

It wasn't until the 1970s and 1980s that we really saw a boom in the market for high-end recreation and competition horses. The proliferation of recreational riders and clinicians fueled the creation of this market.

In the training process today, we have to create work to train the horses, but when I was growing up, our horses were literally being trained every day because we used them for work. We were actually discouraged from specifically "schooling" our horses because that would take away time from working cattle, or whatever else we needed the horses to do that day. The horses would get enough schooling in a regular day's work.

It was more of a concern to conserve that horse for the end of the day, rather than try to make him a better horse in the beginning of the day. It was an unwritten rule that we didn't train on our horses in the morning because if we did, then they were tired when we really needed them later in the day.

If we wanted to take any extra time with a horse, such as getting him used to the bridle or schooling a colt, we usually did that in the evening at cow camp. Most of what we did was on-the-job training. When we had 300 head of cattle to cut out, we didn't have to school our horses on cattle. By the time we had cut out 50 to 75 yearlings, our riding horses had plenty of practice.

We might get on young horses and trot 10 or 20 miles from the cow camp, and then sort and brand some calves, move cattle, etc. A four- to six-hour day would be a short day. In the summer heat we trotted out at 3 a.m., and if we were done by 12 o'clock and there was a job to be done a few miles away, we might go on to that task. We didn't just call it a day.

There was a minimal amount of fencing in those days. We had one camp where there

These two hackamores were my great-grandfather's, Joe Black, and were given to me by Paul Black, my uncle.

A lot of buckaroos make their own gear, and I've followed the tradition. I made all three hackamores and one of the mecates shown here.

was a fence but on only one side of it. We could leave camp in the morning heading south and ride all day without opening a single gate.

A horse can cover eight miles an hour, just walking and trotting at a good pace. When we worked cattle, we traveled the same way, just working back and forth around the cattle. If we were on the move and averaging eight miles an hour and were out there 10 hours, we might cover close to 80 miles, or more, on one horse.

I recently ran across a journal I kept when I was 14. One entry describes a typical 10-hour day: My horse and I left one basin pushing some cattle, and then trotted more than five miles to push another group of cattle, and then went on to another group. When I added it up, I'd probably covered over 30 miles—and that was riding just from one camp to another. It wasn't counting all the back and forth and the jobs my horse and I did along the way.

In those days, most ranches had two types of horses: the "first-string" horses and the tough horses we mainly just used when we needed to cover a lot of country. The first-string horses were the ones we saved for sorting and branding when we needed a better horse. The second-string horses were the ones that might buck or run off. They weren't exactly the horses we wanted to show off to the neighbors, but they were tough, and if we had to go all day and weren't doing anything too technical, they were what we rode.

Starting Young Horses

One thing I see today that concerns me is the intense training that many people are putting on 2-year-olds. When I was younger, we never rode our horses hard every day. I believe in quality, not quantity.

In my opinion, the best plan is to ride a 2-year-old just three or four days a week for a month to two months. Then, when he is progressing nicely and feels good about himself, turn him back out for a month. Let him run and graze and grow up in an

Through the years numerous hackamores and bosals have been collected for various horses.

When I was born, my dad sold a spade-bit horse to a friend to pay the doctor bill. The man bought this bit to ride the horse and 13 years later traded the bit to me to start some colts for him. This was my first spade bit.

This is a bit I made while in high school. By most people's standards it would be considered crude, but most every horse likes the taste and feel of it.

This bit was made by Albert Teitjen in the 1940s, when my granddad bought it for my dad, who gave it to me in the 1990s. After being worn out, the bit was rebuilt to a new shape in 1980.

A bit, made in the 1890s by L. D. Stone, was my great-grandfather Joe's bit and late was used by Paul Black. The bit was a favorite for nearly 100 years, so this copy was made for me, and horses accept it fine.

environment where he can be as much of a horse as God intended him to be. Then after at least 30 days, bring him right back in and pick up his riding again. He is not going to know any more, but he gets more confident in what he does know.

Tom Dorrance always said, "If a horse is going good when you turn him out, he'll come back better. If he's going bad when you turn him out, he'll come back worse."

Unfortunately, I don't think there is enough credit given to this theory today. What usually happens is that people start a 2-year-old and just keep going with him. Often, that horse gets injured and then he has to be laid up as a result. If the last thing the horse remembers from his training is negative, when he heals and comes back into training, this shows. He'll still be "shell-shocked."

The key is to get the horse to the point where he thinks he's invincible and bullet-proof, and then turn him out. That way when he comes back, he knows he's invincible. Even if it's a year before you bring him back, if you did your initial work correctly, all you have to do is get the "fresh" off him and go back to work because the horse is already "broke."

Today, it's certainly possible to ride a 2-year-old just three times a week. So long as he's progressing, you're on the right track.

The problem is that, in most cases, it's more about the business end of things, not what's best for the horse. Some trainers don't want to start a colt and then turn him out because then they can't continue to bill the owner for training. If an owner sends me a 2-year-old and asks if I'm going to ride his horse 30 days a month, this tells me he's more concerned about his money than his horse.

Principal Theories and Truths

You can't learn about wine without tasting wine. You can't learn to taste wine from a book or a video or any other information I can give you. It's the same with horses; the only way you can learn about a horse is from a horse.

You can learn equine anatomy by reading and studying, but you can't learn horsemanship anywhere else but from a horse. The horse is the only one who can tell you if what you're doing is working. If the horse doesn't approve of what you're doing, then it's wrong. You're dealing with an animal that has feelings and emotions. Your own feelings and emotions, and how well you present your message, determine if you pass or fail the test. If your methods and approach are too forceful, the horse reverts to self-preservation mode, and then you don't get anywhere.

"Tom Dorrance said, 'If a horse is going good when you turn him out, he'll come back better. If he's going bad when you turn him out, he'll come back worse.' Unfortunately, I don't think there is enough credit given to this theory today."

After many years, I've realized there are a number of basic truths that apply to every horse, no matter what breed or use. I'll dive deeper into some of these theories in later chapters, but I'd like to introduce them here.

- It takes pressure for relief to be effective, and it takes relief for pressure to be effective, with more emphasis on relief to influence the horse. Sometimes I hear the term "release" used. We might release the physical pressure, but the horse might not acknowledge relief until sometime later. We need to wait until the horse acknowledges relief mentally, not just the physical release.

  A good horseman must be able to read the horse's mindset and his body language. Is the horse seeking relief or giving to pressure? When he seeks relief, the horse searches for a comfortable place to be mentally. When he gives to pressure, he looks for an escape and is not looking ahead; this horse might be intimidated or resentful.

- How and when you apply pressure and give relief determines the horse's attitude and response, and either builds or breaks down his confidence.

  This is why one of my basic training philosophies is always to build the horse's confidence. The more confidence a horse has, the longer he lasts in the training program.

"Is the horse seeking relief or giving to pressure? When he seeks relief, the horse is searching for a comfortable place to be mentally. When he gives to pressure, he looks for an escape."

- Horses don't reason the way we do. We plan for the future; horses look only at the present. This is what makes people greedy, storing up more and better. Horses want only comfort at the present time and respond accordingly.

  Unlike humans, who reason and can experiment with different options and hope for different outcomes, the horse simply relies on what he knows. This is why it's so important to communicate with the horse clearly in ways he understands, and put him in situations where he can respond in a positive manner.

- Horses have incredible memories, no matter if the experience is good or bad. Their decision-making is based on self-preservation, present circumstances, and past experiences.

  The horse looks at a situation, takes an impression from it, and then makes a decision based on his past experience. Some of the worst problems you find in a horse are caused by trauma in past experiences. You can start building the horse's confidence, but as soon as you expose him to real-world situations, he might have a "flashback." No matter how much confidence he gains, it's not the same as working with a mind that is unscarred by trauma. If a horse has been brutalized by a training program in the past, he remains limited—even if he has great physical ability—because he cannot forget what's happened to him.

- If something works with a horse, there must be a reason why, and if it doesn't work, there must be a reason why.

  If you can understand what prepared or caused the horse to react or respond in a certain manner, then you can learn to identify and influence the horse's action in a positive, productive way.

- If you can identify the problem with a horse and the problem's cause, then you can eliminate the cause, and the problem goes away.

  Don't look for a solution to the horse's problem; look for the cause of the problem. Solutions generally glorify or financially benefit someone else although the horse always pays the price. If the cause of the problem remains, the solution doesn't last.

- The horse's confidence level—not his performance level—should serve as a gauge for the rider's expectations.

  A horse can perform out of confusion or fear, but neither is ever a positive thing. I believe in waiting for the horse and finding out what that individual has to offer. Sometimes, all a horse needs is a soft hand and a kind heart.

I'd like to see the old ways preserved. They were good enough for hundreds of years before man put a rush on everything, and, long-term, I still believe time makes a better horse.

"We must be aware that, where the horse is concerned, 'feel' is emotional, mental and physical."

3

Methods of Training and Developing Feel

The greatest horseman in the world can't teach a horse to do anything that horse can't already do on his own in the pasture. Left to his own devices, the horse might never have a reason to do many of the things we ask of him, but the fact remains: He is capable of doing them.

Since we are working with an animal that already has remarkable abilities, the key to success is to be wise enough and have enough skills that we teach the horse in a way he clearly

We can learn a lot about the natural movement of a horse by watching loose horses move without riders inhibiting the horses' action.

The "A" Test

Here's a simple test I often give my students. I start by saying, "I want you to listen carefully and do exactly as I say. Write an 'A' on this paper."

Every student writes the letter "A" on the paper. Some make a lower-case "a" and others make a capital "A." They are usually very confident in their actions.

Using the exact same tone and phrasing, I repeat, "I want you to listen carefully and do exactly as I say. Write an 'A' on this paper."

The students who made a capital letter usually switch to lower-case and vice versa.

Again, using the exact same tone and phrasing, I repeat, "I want you to listen carefully and do exactly as I say. Write an 'A' on this paper."

Every student looks at me blankly and hesitates, completely unsure what to do next.

So I repeat, "I want you to listen carefully and do exactly as I say. Write an 'A' on this paper."

Some students keep trying to write different types of "A," hoping to get the right answer, while others put down their pencils and quit.

I keep repeating the instructions in the same manner and even start asking the students how long they went to school, if they're as smart as they think they are, etc. It doesn't take but a few minutes until every single student is frustrated and confused, and most have totally given up trying. The few, if any, who keep trying show obvious anxiety and uncertainty.

I use this little lesson because it is the perfect illustration of how repetitive push-button training comes across to the horse. The horse goes from willing to confused to frustrated to defeated in a very short time. That's because there's no encouragement that he's headed in the right direction. The reward comes only after he stumbles into the correct response. The horse eventually gets from Point A to Point B, but he doesn't receive any feel or support from the rider that things will get better.

When I keep repeating the instructions over and over, and begin questioning the students' intelligence, it's exactly like the rider who puts the spur on his horse or goes to a bigger bit to be more forceful.

The main problem here is the method of communication. We often try to communicate with the horse in the way that we think, which is as a predator. The horse is wired differently because he's a prey animal. It's no different than if I speak only English and try to communicate with someone who speaks only French. I can ask that person 50 times to do something in English, and although I understand perfectly, he doesn't. But if I use gestures and body language, that person will start to get the message. It's the same when training a horse. He can't do what you ask until he *understands* what you're asking.

Keep this in mind and remember that everything you ask the horse to do is an "A" test.

understands. If we're not smart enough to give him a cue to do one thing, and another cue to do something different, then we can teach our horse to do only so many maneuvers because we can only "program so many buttons" or we get confused.

But if we're more into establishing communication with the horse and developing a "feel" through which we offer him support and encouragement, then the horse doesn't have any trouble figuring out which end to move, or which foot to pivot on. We're the ones who have trouble figuring out how to tell him to do it.

Two Approaches

There are two basic approaches to training a horse: repetition and feel.

With the **repetition approach**, you push the same button, giving the horse a cue, and keep doing it over and over until you get the desired response from the horse. Only then do you give him relief to **reward the results**.

With the **feel approach**, you support and encourage the horse through each step of the process, motivating him through pressure and relief, and **rewarding each "try"** he makes along the way. This allows the horse to make decisions, learn, and understand the purpose of what you're asking.

By far, more people use the repetition method of training, simply because it's easier for the person. It's more mechanical and less helpful for the horse in that it's a more military approach. The rider actually becomes a dictator. Eventually, the horse learns that when the rider gives a specific cue, he's supposed to do a specific maneuver or perform a certain way. This develops a style of "push button" communication, but limits the rider because there are only so many buttons to push.

For example, if you train a horse to do 10 maneuvers, then you need 10 buttons to push. But the truly confusing and frustrating part for the horse is that he doesn't understand the buttons to start with. He only realizes what you're asking him to do after much trial and error when he stumbles onto the correct response.

You're probably asking yourself, "But don't we need repetition in the training process?" The answer is yes, but only after the horse begins building confidence in what you are asking him to do. Once you arrive at this point, you can start repeating your request,

In this photo, I'm showing White Eye in Elko, Nev., in 1983. This horse had as much feel and try as any horse I've ever ridden.

but the critical thing is to provide feel and support to let the horse know he's making the right effort along the way. I compare it to a warm stove on a cold day. The closer the horse gets to doing the right things, the better he should feel.

Obviously, there are some things we do over and over with our horses, but our goal should be more focused on building the horse's confidence than on simply accomplishing an exercise or specific maneuver. We have to build his confidence as he reaches toward the goal—not just after he arrives. If we don't have enough feel to support and encourage the horse's confidence, what we do doesn't have much meaning to the horse, even if he reaches the goal. That's because he just stumbled onto it without using his thought process to get there.

If the horse can feel his way to the target and maybe try some wrong avenues along the way and feel some discomfort, then if he finds comfort when he takes the right avenue, it becomes harder to get him to take the wrong avenues in the future. I want my horse to make decisions—right or wrong. If he makes a wrong decision, I just make things uncomfortable for him, and soon he doesn't want to do anything but the right thing because it feels good. The key is that he wants to do the right thing.

"If he makes a wrong decision, I just make it uncomfortable for him, and soon he doesn't want to do anything but the right thing."

I want the horse to understand what's going on. I don't want him to think, "I don't know why we're doing this, but I know I'll get spurred in the ribs if I don't do it." That kind of training doesn't give the horse any credit for thinking.

Levels of Response

Any time you ask a horse to do something, be aware of the different levels of response and look closely for them. Certain steps take place before the horse gives his full response.

1. **Acknowledgement:** The horse realizes a signal is coming and prepares to respond. This is when it might be beneficial for you to wait and feel for the change in the horse, instead of pushing or dragging him through the maneuver. He might be able to find the correct response on his own. Your goal is always to give as light a signal as possible to gain the desired response.

2. **Preparation to respond:** The horse shifts his weight or body or alters what he was doing in a positive way as he prepares to respond.

3. **Search for escape:** This next level occurs when you don't recognize and encourage the previous two levels as the horse is making a positive effort, even if it's a small effort. If you continue with the pressure without any relief, the horse feels he cannot find a comfortable solution to escape the pressure. At this point, he might begin to push on your hands, or alter his head position, which then alters his balance and further complicates the maneuver. He also starts to become distracted from the task at hand since all he wants is relief. By this point, the horse responds with confusion, fear, anger, resentment or another emotional response—none of which are helpful.

Motivating the Horse

We must be aware that where the horse is concerned, feel is emotional, mental and physical.

For example, when I give the horse a cue, I pay close attention to see if he's interested, scared, frustrated, looking for comfort or resenting pressure. A lot of emotion in the horse goes on during the training process. Whether I apply more or less pressure is determined by his mental-emotional state. If he's already frustrated, applying more pressure is a mistake.

In the physical realm, I must notice if the horse's body is getting into position to do what I'm asking of him. His action starts with a thought before his body ever moves. I feel what the horse is thinking from his body language and how he feels underneath me.

Let's say you want your horse to side-pass to the left. You put your right leg on the horse, but if you haven't set a boundary with your reins, the horse might take three steps forward before you pull back on the reins and stop him.

Let's look at things from the horse's perspective: When he feels leg pressure, he automatically thinks he must move forward to get relief from that pressure.

There are many ways...

...to test the communication level with our horses...

... and see if we can place the feet...

What should have happened: The moment the horse started to move one inch forward, you should have put on the brakes by setting a boundary with the reins, telling the horse not to go forward. Then, when the horse reached toward the left—with any foot, at any angle—the rider should have let up on the gas by giving less leg pressure. This encourages the horse to look for a different route, other than moving forward, to get relief.

You need to "open the door" for the horse so he realizes that moving sideways gives him relief. The important thing to remember is that the horse must feel that he's getting relief or he loses his motivation.

The motivation here is comfort. If the horse starts to move forward and you put more rein on him, he realizes it's increasingly uncomfortable the more he tries to go forward, so he decides to stop going there. He finds it more comfortable to move sideways because you've opened that door to him. He responds to your feel and encouragement.

It helps greatly if you always remember that horses are simple, honest animals. All they want is comfort. The horse isn't looking for trouble. No matter how frustrated you might get, realize that the horse doesn't want to antagonize you or take advantage of you. I've never known a horse to take advantage of a person, but I've seen the person lose the advantage over the horse. You must at all times evaluate the horse's knowledge and where he is in his training, and base your actions on these facts.

Sometimes you get the wrong answer because you're asking in the wrong way. The key to successful training is presentation.

"If the horse finds comfort when he takes the right avenue, it will become harder to get him to take the wrong avenue."

If you can motivate the horse in a way so that the horse interprets the pressure as self-inflicted, what you want makes perfect sense to him. Instead of you putting pressure on him, he feels like he put pressure on himself, and if he understands how to get away from that pressure, he stays off it.

... with the precision we might need...

... to accomplish anything...

... we might ask of them.

Once he understands, a horse doesn't even have to feel actual physical pressure. If the horse simply feels the threat of losing his comfort, he'll move his face, foot, hindquarters or whatever you ask him to move. You just have to present the message to him correctly. Even the best horse can't give you the desired response if you don't clearly communicate what that should be.

Pressure Versus Relief

As we touched on in Chapter 2, a big part of my training philosophy is seeking relief versus giving to pressure.

It's a common belief that the horse's instinct is to go against pressure. I'm not of this opinion. I believe that a horse instinctively moves away from pressure; he tries to avoid pressure in the first place. This relates to the way horses evolved as prey animals. If there are wild horses in an area, your mere presence makes them uncomfortable enough that they leave to avoid the pressure. You don't have to walk up and touch them to make them respond. The horse's first reaction to a threat is to flee, which is essentially going away from pressure. Only when the horse can't escape the predator, or pressure, does he go against it.

So, how do you know if you're motivating the horse with pressure or relief?

"Most people don't realize just how little discomfort it takes to motivate a horse."

If you must use more and more pressure as you progress in your training program, that's because you're motivating with pressure and the horse is getting calloused. If you must use more bit, more spur, more discomfort to motivate the horse, you're motivating with pressure, not relief. When the horse is motivated by pressure, he thinks more about the discomfort and gets out of the way of the discomfort caused by that pressure.

On the other hand, if a horse is motivated by relief, he does all he can to enjoy the experience as much as he can. Horses, unlike people, aren't greedy. This is an important difference to note. People like the nicer house, the prettier car, so they can be "better" than someone else. With horses, this isn't part of their makeup. As long as they're comfortable, they don't worry about tomorrow, or their appearance or anything else. The horse doesn't care what he looks like after a roll in the mud. Horses don't plan ahead to make life better for themselves if they're comfortable right now; that's all they need and life is good. That's why they're often easy to train. If you can make a horse comfortable or uncomfortable, you can motivate him.

Most people don't realize just how little discomfort it takes to motivate a horse.

The horse is so sensitive that a single biting fly can motivate him to travel for miles to get up on a ridge where the wind keeps that fly away. It doesn't take any more discomfort than that to motivate the horse to change his position or actions. If we can learn to apply just enough pressure properly and at the right time so the horse learns to seek comfort and relief, then we only need to apply as much pressure as that lone fly would.

Remember: It takes pressure for relief to be effective, but it also takes relief for pressure to be effective. Any time the horse experiences more pressure than relief, he tries to fight what he can't tolerate, and tolerates what he can't avoid. Neither of these scenarios leads to a thinking, willing partner.

If more of your time is spent putting pressure on the horse in order to get him to respond, you need to ask yourself if he understands, but isn't motivated. Or it could be that he doesn't understand, or isn't ready mentally or physically. You must consider these possibilities carefully so you can achieve success.

About Those Feet

I find there's a lot of emphasis put on the horse's feet. Many people, because they've heard this is important, try to concentrate on where their horse's feet are, but most can't even tell when the horse is starting to go forward or backward.

Feel is actually one of our weakest senses. To develop more feel, we need to learn to be more aware.

We've all heard that a blind person's sense of smell is much more developed than the average sighted person's. Because the blind

Keeping life in the feet helps to maintain a sensitive feel on the reins while side-passing.

Identify and Move a Single Foot

This is a simple exercise, but one that challenges many riders. It helps to have someone on the ground watching, who can confirm that what you think you felt was actually what happened.

Before you start, look down at your horse's feet so you are sure of his foot position. Then sit up straight in the saddle and position your hands and feet to ask your horse to move one end—either front or back—without moving the other. Announce out loud the foot you are going to try to move first so the person on the ground knows what to look for.

Most riders are surprised to find that the horse takes more than one step and also moves the other end of his body. Very few people can make a horse move just one foot because they aren't aware of what the horse is getting prepared to do.

Did your horse move forward? If so, you need to block him from going forward by drawing him back more. Did your horse move backward? If so, you need to push him forward more. You might be using too much leg or not enough. You can't tell what you need more of, or less of, until the horse gets ready to move.

The truth is that you can do this exercise successfully, even on a green colt, if you can prepare him and block him from going in the wrong direction. It's all about how you present the message to the horse and how the horse interprets that message. If he doesn't deliver the response you want, you need to shut the door he's trying to go through, and open the door you want him to go through in order to make it more enticing for him to find the right way.

It all comes back to feel. If you can't feel what the horse is getting ready to do, he's going to do the wrong thing a large percentage of the time.

It might take both my feet to keep the horse from backing, both hands or reins to block forward motion, an inside rein to direct the front end, and an outside leg to bring the hind end—and with every step, I might need to adjust to place each foot where I want it.

person can't rely on his sight, his other senses naturally become more heightened.

Too many people depend on *looking* at the horse's feet instead of *feeling* them. Actually, I think it's much easier to feel how and when the horse's feet move when a person focuses more on the horse's body preparing to move the feet.

The horse must move his body before he moves his feet. The first thing he moves is his withers, to shift the mass of his body, because once the weight is off the mass, he can move his feet. When the saddle horn moves one inch, the horse is getting ready to move. That's the first signal that the horse is going to move his feet. If the rider doesn't change things at that moment, he continues to go in whatever direction the saddle horn is moving. Many horses are headed in the wrong direction before they take even a step, and the riders don't stop them.

If you just wait to see where the body is going to move, you'll be behind in your next cue, or message, to the horse.

Step at a Time

A lot of riders can complete a successful reining pattern on a horse, but aren't able to make him put his foot in one designated spot on the ground. I've seen horses that have won national competitions in reining that you can't move around to open a gate. That horse is programmed for specific maneuvers, but if you try to position him, you can't.

I believe that you should be able to move your horse's feet in any direction you want. I often refer to a horse taking a half step, which is when he just moves one foot at a time and covers half the distance of a full step.

There are many real-life working scenarios where this comes in handy. For example, when I'm roping in the herd, half steps are very important. If my horse takes a full step toward a calf, it might turn and leave. But if I cue the horse to take a half step with his front end and then a half step with his hind end, then I know how many more half steps the horse can take before I can throw my loop without being too far from the calf. When I rope, I work off the cattle's flight zone, and I must be able to control every step. That's not as difficult as it sounds once the horse knows his job.

Another time half steps are essential is when my horse is stepping toward a cow and really reading her flight zone, the area around an animal that makes it move to maintain a safe distance when someone or something approaches. My horse might take a step with one foot and then wait and look at the cow, testing to see if he can take another step. It's almost as if a lion is stalking prey, and that's exactly what I want him to do. Half steps and single steps are really important when the horse is feeling his way to move the cow. I just barely want to move her, not go blasting in.

"Feel is actually one of our weakest senses; to develop more feel, we need to learn to be more aware."

If the horse has enough momentum that he's going to take more than one step, I need to slow him down as soon as he starts to take one step. This relates back to developing feel, as we just discussed.

What about desensitization?

When I was growing up, we never thought of desensitizing a horse. Our whole focus was on directing and controlling the life, the movement and energy of the horse. We had places to go and things to do, so the more life our horses had, the easier it was to get there and get things done. We had to be fairly good riders to stay on some of those horses, but they always had plenty of energy to get the job done.

With the evolution of the recreational, or "hobby" horse, there are many riders who don't ride often or don't ride well. As a result, there's been a focus on bringing down the horse's energy level, and desensitizing has become popular.

I actually see this as taking the energy, life and motivation out of the horse to the point that he isn't even going to move until you tell him to. He tolerates any amount of activity, pressure or pain until you make him move. Once the horse gets to this level of complete tolerance through repetition, he has no feel left. If desensitization has destroyed

Step on the Target

Here's another exercise to help you fine-tune the control of your horse's feet. Pick a spot in the pasture or arena and designate it a "target." It can be a wet spot, a handful of sawdust, a clump of grass, or anything else that won't be intimidating to the horse when you ask him to step on it.

Now, try to make your horse put a specific foot on that target. For example, try to make him put his left front foot directly on the spot.

You'll probably be all over the map to start, but with practice you learn how to feel if the horse is getting ready to reach, or "overshoot the runway." You begin to learn how to rate his movement and slow him down, and that's when those half steps come in handy.

This might sound like nonsense, but it's the same as riding to unlatch a gate, positioning to rope a calf or measuring your position to turn a cow.

As the horse approaches the target...

... each step needs to be measured...

the horse's sense of self-preservation, then you must inflict pain to establish some degree of self-preservation again.

God made horses to be suspicious of humans. When we overpower this and take all the sensitivity out of the horse so he has no feel, then we have to inflict pain by using a crop, spur, stronger bit, etc., to make him respond.

This is one of the things that bothers me most about the training world. There is so much opportunity to offer the horse more feel and a better deal if people would only become more aware. Horses have such sensitivity and feel until we take it from them.

Instead of "dumbing down" the horse to the human's level, I try to make the person more sensitive so he can come up to the horse's level. Of course, it's much easier and faster to just go buy a bigger bit, so that's why some people don't take the time and make the effort to rise to the horse's level.

If the average horse owner goes to the feed store and says to the salesperson, "My horse won't stop. What bit should I get?" the salesperson won't tell him to improve his hands and his horsemanship; he just sells the customer a bit.

Bit companies tell you there's a bit for every problem you have with your horse. This leads to the thinking that if you have a few dozen bits in your tack room, you don't have any more problems. We all know that's not the case.

... and adjusted...

... so the final step is on target.

"Every rider needs to understand and feel how his or her body position, as well as rein aids, helps or hinders the horse's balance and movement."

4

The Power of Position

To be successful at anything he does on the ranch, on the trail or in the arena, your horse must have confidence. The key to developing his self-assurance is eliminating confusion and fear, and creating a positive experience he willingly repeats.

Through the years of developing many cow horses, I've come up with a number of philosophies and techniques for making a confident horse, and for getting a horse and cow to do what I want in the most logical and efficient way possible. I've discovered how to influence the horse's abilities and capitalize on his instinctive behavior and past experiences. I try to analyze situations from the horse's perspective and reason with him, and

Lowering my outside shoulder takes the weight off the inside and allows the horse to move more easily through the turn.

Good body position and proper motivation help us operate with a light feel.

devise ways to use cattle as training aids. The theories and methods I share are related in terms of developing a cow horse, but much of the information can be applied to any discipline, especially when talking about rider position.

For your horse to perform at his peak ability, you need to remain in a positive position, one that enhances your horse's abilities. Being able to feel how your horse reacts to your weight and position is an important and often overlooked aid.

Every rider needs to understand and feel how his or her body position, as well as rein aids, helps or hinders the horse's balance and movement. Then you can make the necessary adjustments to your riding to help your horse reach his full potential.

A Different Way of Thinking

The premise of my training program is to make things as easy and as natural as possible for a horse to understand and implement. For years, I've studied how to motivate a horse by making sure he understands his job; by capitalizing on prior experiences with his dam, in the herd or in the pasture; and by using body position and a soft feel on the reins to encourage him to move freely in the direction I want. Many people interpret a horse's motivation as speed—how fast a horse does what he's told—but that's not true. Motivation is just as much about willingness. Your horse needs to be motivated to do something, and if he's unprepared, confused, nervous or uncomfortable, he probably won't want to do it again.

Confidence is built as you gradually expose the horse to the job you want him to do and relate that work to past experiences, so he has a clear understanding of what you're asking of him. Before your horse was saddled for the first time, he already had experience tracking his mother or the herd, changing leads, turning, stopping and doing other maneuvers. A horse naturally knows how to do these things—until you add a rider's weight to the equation.

Most of the problems people have with their horses relate to head position and lateral movement. For the most part, these problems stem from poor rider position, and the horse compensating to avoid being off-balance. A horse balances with his head and neck. If your weight distribution or rein aids alter your horse's head position, balance and body position the slightest bit, you also affect his foot placement and cause unnatural actions, such as an undesirable head position and a dropped shoulder. That's because the horse must be able to move his head in a natural way in order to move his feet freely. If you have a tight hold on the reins, your horse raises his head and braces against the pressure, instead of lowering his head or tipping his nose to one side and moving his feet in the direction of travel, the way he does in the pasture.

The same muscles that this horse uses to pull our weight also help him maintain his natural head and foot placement.

Understanding Balance

Before you can move on to practicing and perfecting maneuvers with your horse, it's vitally important that you become aware of balance and how it applies to the horse.

Unlike a human, the horse doesn't have arms to maintain his balance, so instead he uses his neck and head. The horse can't place his feet if he can't balance his body.

Since a horse can feel a fly land, the weight of the mecate and hackamore is much more of a signal for him to respond.

The rider's good body position is essential for a horse to do high-performance maneuvers. This shot was taken in Winnemucca, Nev., in 1997.

Many people don't understand how the horse balances himself, and end up relying on bigger bits and gimmicks to control head elevation. What people don't realize is that the horse raises his head because the rider has put the horse off-balance, or because he is trying to escape pain.

I've seen wild horses whirl around and make the most fantastic moves you can imagine. It's amazing how athletic the horse can be without a human on his back. The important thing to realize is that it's not the human's weight that can interfere, but rather that weight in relation to the horse's center of gravity.

The horse learns to compensate for the rider's weight by transferring his own weight to different quarters, and repositioning his head and neck. The rider can either enhance or inhibit the horse's ability to balance himself in motion, which is why it's important to understand the effect of the rider's weight in relation to the horse's center of gravity.

Ask most people to define the horse's center of gravity, and they say it's the center of his back. In truth, his center of gravity is defined by equal gravitational forces in all directions. For example, the horse's center of gravity is different when he goes forward, when he turns, and when he stops.

"A horse balances with his head and neck. If your weight distribution or rein aids alter your horse's head position, balance and body position the slightest bit, you also affect his foot placement and cause unnatural actions."

It's more critical for the rider to help the horse stay balanced in a turn or when stopping than when the horse is traveling forward in a straight line. The horse doesn't have trouble balancing himself if the rider's

It's obvious from my left hand that the horse isn't pulling hard, but I notice the nose and poll aren't vertical. This means that although the horse's neck is bent, he's still showing resistance through the poll.

Here, the poll and nose are vertical, which tells me the poll has softened.

weight is behind the center of gravity; this is actually how most cutters ride.

Let's look at the horse about to perform a sliding stop. In this situation, it's definitely better for the rider to be slightly behind the horse's center of gravity than to be ahead of it, or shifting back and forth. Because the horse's shoulders are elevated in the sliding stop, if the rider's weight comes forward, the horse must elevate his head in an effort to regain his balance and shift the center of gravity back. This is why it's so important to maintain position throughout the stop.

Be aware of which muscles you tighten as the horse prepares to stop. If you sit up straight and tighten your back muscles, your horse's power and momentum pull you forward, putting you ahead of his center of gravity. But if you round your back and tighten your chest and abdominal muscles, this pulls you down, lowering your center of gravity and helping you stay behind the horse's center of gravity.

Watch Your Weight

When I was younger, an excellent bridle horseman in Nevada told me that the key to moving a horse to the left is to put more weight on his right side, and vice versa. I grew up with this philosophy and, as I got older and tested it on thousands of horses, I discovered that his theory is sound and applies to other things we do with our horses. I've found plenty of people who might disagree, but I haven't found one horse that does.

It's fairly simple for your horse to remain balanced on a straight line. Even if you

As my weight shifts outside the horse's center of gravity, the horse contracts the inside muscles along his spine to pull himself in an arc.

If my weight is inside the horse's center of gravity, he tends to drop his shoulder and/or straighten his spine by contracting the muscles on the outside of the turn.

shift your weight to the front, back, left or right, he usually can compensate for your changed position without altering his path of travel. Most balance-related issues occur while a horse is changing direction or doing a lateral movement. These movements are the foundation of just about anything you do with a horse, including lead changes and turnarounds. If you want to maximize your horse's potential in any discipline, you must learn to position your body so that it helps, rather than hinders, his movement. You can eliminate several training issues by simply staying out of your horse's way.

Think of your horse's spinal column as a chain with his poll and loins on either end. If you grasp the end of the chain that represents your horse's nose, and drag it to the left, all of the links, or vertebrae, follow the same path in an arced manner.

Now, watch your horse change direction in his pasture. He first tips his nose in the direction he wants to go, and then his feet and the rest of his body follow. If he turns left, it's natural for his body, from poll to loin, to be arced in the direction of the turn. His sacrum, the wedge-shaped bone that lies below the lumbar vertebrae and makes up the pelvic wall, is straight and doesn't bend.

Basically, as the horse turns to the left, the muscles on the left side of his body contract to help him stay balanced and pull him through the turn. The muscles on the right are relaxed. The horse has to disengage the muscles on one side before he can engage the muscles on the other side. When you add the weight of a saddle and your body to his back, you must learn how to remain positioned to the outside of your horse's center of gravity, at the point where there are equal gravitational forces in all directions, to avoid impeding his movement.

To understand center of gravity, think about skiing or riding a four-wheeler. You must place more weight on the outside ski or the outside of the vehicle to lighten the load on the inside in order to turn smoothly.

The same is true with your horse. If you ride him in a circle to the left, but distribute more weight on the inside (left) stirrup than on the outside (right) stirrup, you push him in the direction he's already pulling himself. He has to contract the muscles on his right side, which should be more relaxed, to pull himself, and you, through the turn without getting off-balance.

Understanding Balance

To better understand balance, try this simple exercise: Sitting on your horse at a standstill, shift and put more weight in the right stirrup, enough to pull the center of your saddle off the horse's center. Walk the horse in a circle to the left. Then quickly change directions. Change to shift your weight to the left and walk the horse in a circle to the right. Do this a few times so you and the horse get a feel for the changes.

Now, do just the opposite. Put more weight in the left stirrup and walk the horse in a circle to left. Watch what the horse does: He drops that left shoulder and his head comes up. He's trying to compensate because you've put him off-balance. This is especially obvious in a young horse that hasn't yet learned to pack a rider's weight.

You often see a horse that leans to the inside, or drops his inside shoulder. That's usually the result of the rider leaning to the inside, which adds weight to the horse's front end, disengaging his hindquarters. Furthermore, the horse tips his head to the outside to help stay balanced. The rider applies firmer, direct-rein pressure to pull the horse's head into the turn, which causes the horse to brace and resist the pressure.

I see this frequently in amateur cutting competitions. A rider anticipates the cow's movement and leans in the direction of the cow. The horse stops, throws up his head and disconnects his focus from the cow, because he can't move agilely when the rider weighs him down on the inside.

On the other hand, if you place a little more weight in the outside stirrup before and during the turn, your horse's inside muscles can pull him through the turn. And, without the interference of your weight, he works off his hindquarters, which lightens his front end so he can move freely and remain balanced.

These principles also apply to backing your horse and encouraging him to move forward. I've found that it's more natural for a horse to back if I shift my weight slightly forward, removing the pressure from his hindquarters.

When riding a green horse that's hesitant to transition to a trot or lope, I've discovered that if I shift my weight backward and sit deeply in the saddle seat, the horse moves forward freely and willingly. This indicates that having weight on the horse's front end blocks forward movement, and when that obstacle is removed, it frees the horse's front end so he can drive himself forward from the hindquarters.

You can feel this if you sit on an exercise ball. If you lean forward, the ball pushes backward. If you lean backward, it pushes forward. And, if you lean to one side, the ball pushes in the opposite direction.

However, a horse can learn to compensate for and push the rider who is inside the center of gravity. There are world champion riders who ride to the inside, proving that the horse can learn to work this way. But it's the opposite of what a horse does naturally, and he must go through a "rehab" program to learn this new way of thinking and moving. I always try to consider what's easiest and most natural for the horse, and if I can remove my weight as an obstacle, then I don't have to put him through "rehab."

As I ask the horse to back, putting more weight on my thighs and less on the horse's loin allows him to back easily. This is emphasized when backing a horse up a slope.

"We must develop judgment so we can compensate in every circumstance; this is particularly true when we add livestock to the equation."

5

STOCK-HANDLING MANEUVERS

When it comes to working cattle, your horse must rely on a variety of basic maneuvers countless times. You also use a number of these maneuvers to get your horse in position to rope. Although these are all basic maneuvers the horse can easily do on his own when loose at pasture, you gradually fine-tune them as the horse progresses in his training.

A horse has experience and knowledge at rating before I ever ride him. That helps me if I can tap into that and transfer it to the cow. This horse's expression shows that he's on the balance point and in control of the cow.

Even after you have mastered a maneuver, you must remember that nothing is static with horses. You shouldn't simply develop certain maneuvers and then apply them in the same way to every situation when working cattle. You must develop judgment so you can compensate in every circumstance. This is particularly true when you add livestock to the equation.

"The horse has had miles and miles of experience with rating before I ever saddle him."

Your horse stops, backs, turns or pivots frequently when working cattle, but those maneuvers can differ greatly from the same basic moves done in an arena with no cow in sight. For example, a horse in a reining pattern might back rapidly for many strides, while a horse working a steer might back only a couple steps before pivoting in another direction. You need to trust the horse and give him leeway to use the maneuvers efficiently to accomplish the task at hand.

This chapter looks at a number of basic maneuvers you use when working livestock and covers some exercises that help you add precision and smoothness.

Rating

Rating might not be defined as an actual maneuver, but it is a natural instinct of the horse and one that I can use to my advantage.

Before I ever put a horse on a cow, I want that horse to learn to "rate" me as his rider. What I mean by rating is that I want the energy of the horse to be compatible and match my energy. If the horse is in tune with and rating my energy, he is in a much better position to respond quickly than if his energy is higher or lower than mine.

The horse should learn to take note when my body energy speeds up or slows down and try to be in harmony with me. My goal is to have the horse confident to stay with the life, or energy, in my body.

When I ride a young horse or an unfamiliar horse that doesn't really feel the life in my body, I bring up my energy and gauge his response. If the horse doesn't respond by bringing up his energy, I do what I have to do to energize him.

On the other hand, if the horse's energy is higher than mine and he's too fast, I slow down my energy. If he doesn't acknowledge this and still wants to go fast, that's when doubling can come in handy. (At the end of this chapter, see the doubling section.) I just want to make an impression on him so he doesn't keep challenging me. It's like telling a child not to touch a hot stove: "That's hot, don't touch it!" But if the child goes ahead and touches it, the next time he hears the word "hot," he pays attention.

Of course, I don't want to slow down my energy in preparation to stop and then suddenly jerk the horse into the ground on the next stride. If a green horse just acknowledges the change in my energy and slows down for a stride or two, this is a try on his part, and I build on it.

I don't use my spurs except to reinforce what my legs and body are already telling the horse. Doing otherwise would be like disciplining a child for not cleaning his room before he's been told to do it. He first has to be told to clean it, and then if he ignores that, can be disciplined. If I discipline my horse first, he doesn't understand why I'm scolding him, and this leads to resentment.

Once the horse knows how to rate my energy, I don't have to do much pulling and spurring to get him to do anything. When it comes to rating a cow, I can support the horse more without having to be so abrupt with my hand and legs.

My take on rating the cow is that the horse has miles and miles of experience at rating before I ever saddle him. In fact, one of the first things a foal learns to do is rate his mother and get in beside the mare's flank for security. A colt born out in a field rates his mother shortly after birth. If he has been in a normal herd environment, he has been rating other horses his whole life.

The horse's rating instinct is very keen before I ever put a saddle on him, so I want to capitalize on all this previous experience and have him apply it to the cow. As a foal, he instinctively knew that his mother meant security. My goal is to have the horse make the cow his sense of security. This is actually accomplished rather easily.

All I have to do is make the horse work harder if he doesn't rate the cow, and he

Training Know-How

As the horse learns and progresses through his training, and better understands his job, the less you should have to do. You need patience and understanding with a horse that is lacking confidence, yet you must be careful and, at the same time, effective with a horse that doesn't have motivation or desire. Sometimes an insecure or inexperienced horse requires light, steady contact to support the suggestion of completing a maneuver, but beware of too much pressure from the reins, as this can be counterproductive. Too much pressure with the reins or other cues, and the signal can quickly go from being a respected signal to a resented signal.

The quality of a horse greatly depends on the timing and feel of the hands on the reins. You can use different bits, hackamores or other devices tied at certain angles to develop a headset or specific response, but it really comes down to how you present and maintain feel with your hands. Many of the problems people have in handling horses are a result of feel and timing not being compatible to the horse. This can lead to problems with head position and can make a horse hard-mouthed.

A common problem occurs when you hold steady, heavy pressure on the reins throughout a complete maneuver. The only time you can pull or signal a horse with reins and expect a response is when the horse's feet are on the ground or a foot is just leaving the ground. If you pull as the horse's foot comes down, he must "wear" the pull until his foot is on the ground again and he can reposition it for change. When you make too heavy a pull on the reins, the horse quickly learns to push against your hands. Light contact on the reins can help support the horse's thoughts for a maneuver, but if rein pressure is too heavy and the horse pushes on your hands, you then alter his weight distribution and his footfall.

It is so important to give slack or release of pressure on the reins when the horse can comply. This means you need to get in time with his feet so you apply pressure only when the horse can respond and comply with what you're asking. The horse will position his feet or body to avoid getting his mouth pulled, but only if he understands what you're asking.

When a horse's feet are in the air, I can easily shift his balance and redirect him with a quick pull.

Flight Zone and Balance Point: An Introduction

I go into much greater detail about these two definitions in later chapters, but a basic introduction is helpful at this point.

Flight zone is the area around an animal where, when something approaches too closely, the animal feels the need to move to maintain a safe distance. Both horses and cattle have flight zones. The deeper (closer) you are to the animal, the faster that animal moves. The farther you move out of the flight zone, the more you can slow down the animal or even stop it.

Balance point is the position from which you can influence an animal to change direction or stop. The balance point constantly changes, depending on the position of your horse on the cow. When your horse enters a cow's flight zone and encourages her to move, she chooses a direction in which to go. If you block that path, she usually turns her head in the direction of another route. That's your signal that the balance point is shifting.

Part of the pecking order for every horse is learning the flight zone and balance point of the other horses in the herd. If he doesn't understand, he learns very quickly! The dominant horse always establishes a flight zone in the other horses.

quickly realizes that when he rates the cow properly, he doesn't have to work. Every time the cow changes direction, the horse must change direction. He soon understands that as long as he follows the cow, he doesn't get redirected and he works less. For this reason, the cow quickly becomes a sense of security.

Most horses quickly learn how to rate a cow. I often start out trotting or slow loping, and by the time the cow is getting tired, the horse has it pretty much figured out. Some horses might take a couple cows, but often it just takes one for the horse to learn this basic lesson.

Stopping

There are various ways to develop a horse's stop, but you must develop the idea of the stop before you can refine it to a high-performance stop, such as a sliding stop or a cow-horse stop.

I never want to develop a sliding stop in the horse and then take that stop and go work a cow. I like to develop the horse to

This horse is in the flight zone and moving the cow, and the horse's ears indicate he is looking for the balance point.

work the cow and then everything my horse does—including his stop—is initiated by the cow. It's a stop that's natural for the horse, not necessarily a stop that's stylish for the show pen. I don't want to tell the horse how to be a horse if the horse knows how to do his job. When the horse rates me and rates the cow, the stop comes naturally. In competitive cutting, it's all about the stop, but the horse isn't stopping with the cow all of the time; he's stopping with the rider.

The way I see it, the horse has far more experience stopping when he needs to stop and preparing himself for the next move than he'd gain from whatever I would ask him to do on a cow. A horse practices all these things in the field when he runs and plays. He knows full well how to balance himself and prepare for any maneuver I need him to do. All I need to do is bring out his ability when and where I want him to use it.

When you ask the horse to stop, keep this in mind: When you pull hard enough to make the horse engage his muscles to resist, you affect his balance and natural way of moving and operating. Ideally, you should have a light pull on the reins and keep your weight close to the center of gravity going into and throughout the stop. This allows the horse to perform with the least distraction and find his balance to stop properly.

Although you don't want the horse fighting or resisting the stop, you also don't want him to overflex; otherwise, he gets too much bend and he transfers weight to his front end. When that happens, he must compensate for having his head and neck in an unnatural position, as discussed in Chapter 4.

Stopping Exercise

I use the following exercise to help an older horse that doesn't put out much effort, and also to help a young or inexperienced horse that doesn't know to stop on cue. It's best to do this exercise with a little speed, such as a trot or a slow lope. I want the horse to work a little but not be traveling so fast that he gets confused when I ask him to stop.

As he learns that the pressure is consistent and that the relief also is consistent, he starts to avoid the pressure area and strives to stay in the relief area offered. The correct balance of pressure and relief makes this exercise effective, and leads to a confident, compliant and motivated horse. Too much relief, and the horse becomes lazy, resentful or bored. Too much pressure creates a nervous, angry, frustrated or confused horse that tends to overanticipate maneuvers.

Practice this exercise in an area with good footing, and make sure you have enough room that you can make a complete circle every 10 seconds or so. That amount of time allows the horse to relax between maneuvers, but not lose his train of thought.

"When a colt is born out in a field, he rates his mother shortly after birth. If he has been in a normal herd environment, he has been rating other horses his whole life."

Moving at a trot or slow lope, make a circle to the left. Pick a spot to stop and then roll your horse back to the right. Now make a circle to the right and come back to the same spot. Stop and then roll back to the left. Make another left circle, and continue.

Each circle should be on top of the last circle you made, and each stop should be in the same spot. As you ask your horse to stop, you should cease your body rhythm and just take the slack out of your reins. You can say "whoa" if you like, but the horse should mainly be reading your body language.

Try to work at a speed that, with just a light pull on the reins, has your horse down to a walk in about four steps. One or two steps before the horse slows to a walk, pull one rein just hard enough to elevate his head, but not hard enough to cause panic. When a horse panics, he responds without thinking. The idea of elevating his head is that doing so allows him to acknowledge the discomfort and motivates him to look for a more comfortable deal—which is the stop.

This part is the key to success: The rider offers a soft feel with the hands and allows the horse to position his head where it needs to be to balance him and to prepare to position his feet for the stop. Once the horse understands what he is supposed to do, the discomfort of a harder pull with one rein can be used to encourage more try. When this starts to work, the reward of the short rest develops the confidence the horse needs to think about doing the right thing.

Don't look for a certain level of results to reward the horse. Look for the slightest try, the slightest change. Build on that as the horse's confidence allows.

Done properly, it should only take a few times to make a slight change, with your horse putting more effort into the stop. When you see this change, don't do the rollback; just stop and let the horse relax. Put your

As we lope a nice, relaxed circle to the left, I pick a point…

… and ask the horse for some effort in his stop.

Without hesitation, we make a half-turn to the right and lope …

… into a nice, relaxed circle to the right.

hands on his neck and lengthen your reins so the horse knows he can quit for that session. At this point, the horse may take a deep breath and/or lick his lips. These are positive signs that show he fully realizes the relief. This motivates him to put effort into the stop so he can get this relief again.

Reward all the efforts the horse makes and do not prepare to pull on the reins. After the horse makes two or three slight improvements, call it a day and quit on that session. You can lose progress if you push and get hung up looking for a certain level of performance in one session.

At the same point, I again ask the horse for some effort in the stop...

... and without hesitation we turn left...

... and lope again ...

... and continue stopping and turning until a change is felt.

When the horse puts a little effort in the stop, I let him rest and think about it. After a spell, I can repeat the stop to get another try, and that's all I need—a try, not results. Looking for results can destroy the try, but looking for a try can build the results.

Backing

If you watch a group of horses at pasture, you quickly realize they don't spend much time backing up. The vast majority of their movement is forward motion. A horse at liberty might stop, shift his weight back and turn away from something, but he doesn't usually keep backing away from something. As a result, when you want to teach a horse to back straight and for any extended length you must understand this is not a natural movement for him.

If you've ever tried to push a chain, you know it's not easy, nor is it pretty. The links tend to go in every direction. Yet when some riders try to back their horses, it looks like they are trying to push a chain. The horse's head may be up or it might be tucked into his chest. His shoulders might be going left or right, while his hips are going the opposite direction, and he's definitely not going straight.

On the other hand, if you pull a chain, the links stay neatly in the same line. This is your goal when backing your horse. His spine should be straight from the tip of the tail to the poll, and his hindquarters should be engaged so that he's "pulling the chain" before you—the rider—put too much pressure on his head to "push the chain."

To understand the difference, it helps if you watch someone else back his horse. If the horse's hind feet start moving first, the hind end is actually pulling the front end back. This horse is "pulling the chain." If the front feet start moving before the hind feet, the front end is pushing the hind end, or "pushing the chain."

If the horse's hind foot leaves the ground before the opposite front foot, the hind end is pulling. If these two feet leave the ground simultaneously, the hind foot causes no resistance. If the front foot moves before the opposite hind foot, you have resistance.

It's easier to move a horse backward if he is light and supple and you can move his hindquarters from side to side. Put just enough pressure on the reins that the horse acknowledges you. That might be the slightest movement of his head up or down, looking back, to the side, or anything to show his acknowledgement. Next, put one leg on the side to move his hindquarters over. Use one rein to reinforce only if necessary. As soon as his hind foot moves over one step, cue him to move his hindquarters back over the other way, again just one step. All you're doing is moving the hind end left and right one step from side to side.

If the horse tries to move forward while you do this, be firm enough with the reins to put him back in his original tracks, and then just resume light pressure.

If the reins are not offering enough feel and the horse hasn't stepped back, add just a little more rein pressure to encourage him to shift his weight back. When he does take a step back, you should sit still for a moment, or at least as long as it took for the horse to take the step back. Two or three minutes wouldn't be too long. This lets the horse realize how he got out of the situation.

Your horse needs to back in many different scenarios when working cattle. Just remember: You want to get those hind feet moving first. With a little practice and by not violating the horse's confidence, you can use this lesson to get a colt to take that first backward step, and also to polish the back on an older horse.

Backing Tip

When cueing your horse to back up, it helps if you take all the weight out of the seat, and rise up with your weight more on your stirrups and thighs. When the horse's feet move, he raises his loin and you taking your weight out of the seat makes it easier for him to position and use his hindquarters.

Turning or Pivoting

A horse must frequently turn and pivot when working cattle. To understand what type of turn is most natural to the horse, watch a group of horses at play.

First, let's define what we mean by "pivot foot." This is the foot bearing the most weight for the longest period of time in a turn.

If a horse needs to be fast while turning, he pivots on his inside hind foot. Forward motion is always faster than reverse motion for a horse, so if he's going to turn and wants to be quick about it, he pivots on the inside foot so he can maintain forward motion on the front end.

This is a photo of Popcycle when I was starting him. Note the foot position as he backs. His right hind has stepped back, his left front is mid-stride, and the weight is leaving the left hind. He later developed into one of the most natural stopping horses I've experienced because it was really easy for him to pull with his hindquarters.

What's confusing for the horse is when the rider cues him to pivot on the outside hind foot, and still expects the horse to turn fast with his front end. This is actually a way of backing the horse into a turn, and definitely causes him to lose momentum and slows him down.

I've sat at the National Cutting Horse Association Finals for several years and specifically watched how the horses turned. During warm-up, when the rider directs the horse in how to turn, all the turns were made on the outside pivot foot. Yet over and over again, when the horses were actually competing and the riders couldn't rein them, those horses made the vast majority of turns on the inside pivot foot.

I don't train my horses to turn on the outside pivot foot, but a lot of times, when a horse works a cow, he does it on his own, whenever he realizes that he needs to get out of the cow's way. I don't make an exercise out of backing the horse around and pivoting on his outside foot, although I know this is a popular exercise for cutters.

The idea of backing a horse into a turn on that outside pivot foot is to enable him get farther out of the cow's flight zone. A cutter, for example, moves farther away from the cow in the turn, so he, in essence, slows down the cow. In this situation, the horse doesn't need to be fast because the cutter slows down the cow as the horse, while making the turn, gets farther out of her flight zone. The outside pivot foot gives better position in the turn to get away from a cow, and the inside pivot foot gives more speed in the turn.

Doubling

Doubling allows you to control your horse's speed. It's not a maneuver, per se, but is a part of basic horsemanship. If you can't regulate the speed on your horse, you can't effectively control your horse. Although you don't see doubling in a reining or dressage pattern, a cow horse event or any other advanced performance competition, doubling is an exercise that can prevent a lot of mishaps and undesirable patterns.

Many people aren't clear on the definition of this term. I don't think you can look up "doubling" in a dictionary and define it the way I do, or the way many cowboys do. People have different definitions of doubling, but this is mine: When you double a horse, his hind end quits driving.

With a young horse, I always make sure he understands doubling before I leave the round pen. This keeps me in control and, most importantly, maintains safety for both the horse and me.

To double the horse, take the slack out of one rein as he moves forward. As the slack is drawn out of the rein, the horse should tip his nose toward the tighter rein, with a slight bend in the poll.

Let's say you have tightened the left rein, and the right rein has more slack. When the horse breaks in the poll and in the loin, if his nose is pointed to some degree to the left, the hind feet should step to the right. When a horse can do this consistently and with some confidence, his inside hind foot reaches up and underneath him, toward the center of the girth. The inside front foot takes the shortest steps, just pivoting in place while the other three feet maintain forward motion.

The hindquarters don't have to step out to the side to double the horse. When the loin and poll bend, the hindquarters quit driving. If a horse isn't very controllable or becomes really scared and out of control, that soft feel might not save your life. If you find yourself with more speed and no control, then it's a matter of tipping your horse off-balance as he runs or bucks, doubling him. This takes perfect timing and the right amount of pressure on the rein. This is a procedure that hopefully few people have the opportunity to practice.

However, when all else fails, take the slack out of one rein and hold lightly until the front feet leave the ground and the hind feet are getting ready to. At this point, give a pull, hard enough that the horse recognizes it, while all his feet are off the ground. Then he can't brace against you, and the move takes the horse off-balance. His head goes one way while his hindquarters go the other. When his feet hit the ground, if the rein is still tight, the horse still can drive with his hindquarters and brace against your hand with the rein.

If yours is a smaller horse and your weight shifts to the outside too much, the horse can lose his balance and fall. You don't want that to happen, but the instant the horse finds his balance questionable, he no longer wants to speed up or get elevation by bucking. So just make a quick pull to take him off-balance and then release him for a stride or more until he drives again. Then take him off-balance again. Usually, if this is timed properly, after a few times—at the most—a horse doesn't feel good about being off-balance every time he wants to drive with his hindquarters. He quits driving and looks for something else to do.

Once the horse knows how to step his hind feet out to the side at any speed, he can be doubled. Because the feet are reaching forward, the horse can maintain his balance. The slower the horse goes, the farther he can reach, but at any speed he can disengage, quit pushing with the hindquarters and shift them outside of the line of travel.

Because you get the horse to quit driving by taking up one rein, you can double a horse going down a cow trail and not get off that narrow path. He can stay on the line of the trail, traveling with just a slight bend in his body and without driving with his hind end.

Doubling isn't only about getting the hind feet to quit driving. Sometimes people pull the rein, and the hind feet step in the opposite direction, but the poll is stiff and the horse pushes against the hands. When his poll is straight and stiff, the loin is straight. Then, although the hind feet step out, the horse might instead be turning on the outside front foot. When the left rein is pulled, the hind feet reach to the right, but he pivots on the right front foot. This is a reverse motion, rather than a forward motion, so the left front reaches back. The problem: If the horse panics going forward and only knows how to step out in a reverse motion, he has too much speed to make that transition and disengage his hindquarters.

But if the horse is soft through his poll and loin, he can disengage his hindquarters at any speed and be doubled. Because the feet reach forward, the horse can maintain his balance. The slower the horse goes, the farther he can reach, but at any speed he can disengage, quit pushing with his hindquarters and shift them outside of the line of travel.

There's definitely a timing factor and an art to doubling. When done properly, it can prevent a horse from bucking or running off. With a slightly different presentation, doubling can help prepare the horse for a sliding stop by rocking his weight back and beyond the point where the horse quits driving himself forward and starts pulling himself into the stop with the hindquarters.

Doubling Tip

One of the most important things to look for when doubling is suppleness in the poll. If the horse isn't supple in the poll, he isn't supple in the loin and he's braced mentally. All three—poll, loin, mental status—work together simultaneously, and no training device can achieve this suppleness until the horse is ready to give it.

Most horses give with very little challenge if you haven't given them a reason to be defensive. A horse can bend his neck until his nose touches the saddle, but if his poll is straight, then his loin also is straight. Mentally, this horse is on guard and defensive. He's keeping an eye on a straight path so he has an escape route.

If the horse's poll and loin are straight, his shoulder might fall to the inside or outside. Either way, this can be dangerous because he's off-balance. The horse pushing forward with his hind feet extended more than normal puts more weight on his front feet. If he stumbles in front, he has a hard time recovering because his hind feet are extended behind his center of gravity.

In most defensive maneuvers a horse uses—running, bucking, rearing—his hindquarters are engaged and locked, and his loin is straight. Remember that the horse's poll, loin and mind all work together simultaneously.

A horse can be traveling straight on a cow trail and with the slightest bend in his body quit driving and stop with one rein.

"Ears are a big indicator of what a cow is thinking; a cow's ears really just amplify where her eyes are looking."

6

Cattle 101: Learning to Read and Control Cows

It's been said that, "Wild cowboys make wild cows and quiet cowboys make quiet cows." I think this is true, regardless of the breed of cattle.

You can take cattle with a reputation for being wild or fast and, just like horses, if you give the cattle experiences where you can control their environment and the way they're handled, then you can desensitize them so they aren't as wild. It's no different than handling mustangs. You don't take away all their

As my horse and I approach the herd, I consider how to avoid the flight zones of cattle that don't need to be moved and how to minimize our effect on the flight zones of the cattle around those that we need to move.

wild traits, but if they aren't threatened those traits don't come out.

Cattle usually don't lope anywhere unless their self-preservation instinct is aroused. But if they have been handled fast or roughly, when they see someone approach, their sense of self-preservation is engaged, and they might run off or try to find somewhere to hide.

Cattle Breeds and Types

The breed of cattle you deal with tells you a lot about the disposition. Without going into detail about specific breeds, cattle that are bred to produce meat or milk are usually docile and tolerant, even lazy. With most beef and dairy cattle, a certain amount of self-preservation has been bred out of them through the years because it's not a necessary trait and can actually get in a person's way when working with them.

"A cow's thoughts go in the same direction as her ears."

Cattle that are bred for survivability in harsh conditions, such as Corrientes and Longhorns, can have an entirely different disposition. The flight zone is bigger, and they may be a little smarter or more cautious than the other breeds.

Brahmas are by far the smartest of all cattle. I'd compare their sensitivity more to that of a horse. If you do right by Brahmas, they're easier to train and work with than most any other breed, but if you violate them or take advantage, they can be more aggressive and harder to handle.

You can make some basic assumptions within certain parameters based on a particular breed, but there are definitely personality differences within any breed and within any herd. Always some individuals are more sensitive, and others are more tolerant.

Going back to the statement about wild cowboys making wild cows, how cattle are raised makes a huge difference. Brahma or Corriente cattle, which under some conditions might get pretty wild, can be as docile as any cattle when they aren't taken advantage of or pressured.

Read and Control Cattle

You can't exactly sign up for a seminar or weeklong course to learn how to read cattle. It takes experience and intuition; there is no step-by-step formula. A top hand can share some pointers to help you recognize when a cow is about to move and where she'll go, but you learn best through practice and studying your mistakes.

The good news is that cattle pretty much tell you everything they're going to do before they do it, whether you're horseback or on foot. They don't plan a surprise attack or try to fool you.

A cow tells you the direction she's going, sometimes long before she actually moves a foot. If she stands still, watch her withers area because she starts moving her body before she ever moves her foot. If she makes just a subtle shift forward in her weight, she's going to keep her feet underneath her and go forward.

If a cow is going to move left, she breaks at the poll and tips her head to the left, looking to the left. She does exactly the opposite if she's going to move to the right. Even if she runs, the first thing she does is bring her nose in the direction she wants to go, and her body then follows. It's a dead giveaway: Wherever her nose points is the direction she's going to try to go.

Ears are a big indicator of what a cow is thinking. A cow's ears really just amplify where her eyes are looking. The eyes and ears tell the same story. At times when I have to work dark cattle in poor lighting conditions, it's much easier to watch their ears than to see their eyes.

If a cow's ears are forward, this shows the cow's focus. Her thoughts go in the same direction as her ears. It's easier to draw a cow forward if her ears are forward. If her ears are back, her focus is not forward, so it is harder to draw her into moving forward.

You can prepare and train your cattle just as you would your horse. Cattle want to be in the herd for security; there's safety in numbers. But if you make the herd uncomfortable for a cow, she looks for somewhere else to go and find security. All you have to do is give her a few experiences cutting her out of the herd, and she learns that if she's uncomfortable in the herd, she just has to go somewhere else.

Let's say you're trying to drive a single cow out of the herd. She looks back at you, then to the left and the right, but you keep positioning your horse in those directions so she can only go forward. Then, when you stop following her and let her feel relief from the pressure, the ears go forward like she's looking for that warm stove on a cold day. She can feel the relief, but she doesn't know where it's coming from, and her ears go forward looking for more relief.

You've made it difficult for her to go anywhere else, and if you just give her a little time, she quickly figures it out. You see her ears go forward and she moves right out of the herd—just like you want her to do.

But what often happens is that people try to push the cow too much. Even when she goes in the right direction, which, in this case, is out of the herd, she feels constant pressure. The cow that is pushed too much never learns to leave when she's asked. Next time you cut her out, you have to chase her out the same way. The secret is to give her relief as she's going the direction you want her to go. This goes back to the same thing we consider with horses: Is she giving to pressure or is she seeking relief?

While working cattle on the open range in their natural environment, I spent a lot of time studying how my horse's position in relation to a cow influenced the way the cow reacted and moved. As a result, I found that, for me, one of the best training tools is the cow.

We worked our cattle all the time outside in the open, so for some it was an annual experience to be cut out of a herd individually, or in cow-calf pairs. When we worked cattle in these conditions, they quickly learned when we blocked them to go in the other direction. They learned this process early on as yearlings before they ever had calves, so being redirected wasn't new to them when they were older.

In those days, our cattle might see a rider once every few months, so when they did see someone, they took off in a trot, but we used this to our advantage. If we needed to go five miles, we'd just take off and go. The secret was to make sure the cattle were headed in the right direction, but once we stopped them, they were broke to just hold and stand. It wasn't that we were always mounted on top cow horses either.

You can actually train your cattle and your horse at the same time, but you might be compromising what you get done with the cattle to train your horse, and vice versa. By trying to maintain some balance, you can get them all trained at the same time. You don't need a top bridle horse to work

This cow's ears are forward and focused on me.

The cow's ears are back, telling me that her thoughts are on me.

As the cow turns away from me, I stop my horse, removing the pressure, and her ears go forward, looking for the feel of relief.

some cattle. Still, if you have only one shot to move a group of cattle you've spent all week gathering, there's a big storm coming and you might lose them up the canyon if they scatter, that's not the time to train your horse.

Flight Zone and Balance Point

The flight zone and balance point are often misunderstood, and this can make things much more complicated than they have to be. We touched on the definitions earlier, but let's look at these in more detail now.

"We use the cow's flight zone to get her to move, and we use her balance point to direct her."

The flight zone is the area around an animal where, when something or someone approaches too closely, the animal feels the need to move to maintain a safe distance. Both horses and cattle have flight zones. If that zone is penetrated, it causes the animal to engage its self-preservation mechanism and either flee or fight. Prey animals also seek security from what they perceive to be a potential threat by moving with the herd or, if they're young, near their mothers.

If you've ever watched a good border collie work cattle or sheep, you probably noticed that the dog runs toward the stock until the dog senses the stock's flight zone. Then the dog turns hard and circles around the stock on the outer edge of the flight zone. When the dog reaches the balance point, he carefully moves closer to the stock, calculating the effect of penetrating the flight zone, to get the stock to move. Likewise, the dog retreats to get the stock to slow down or stop. This is a good example of the flight-zone principle.

The balance point is the position from which you can influence an animal to change direction or stop. Picture the border collie as he moves into the cattle's flight zone. The dog enters the flight zone on the balance point—the point at which the stock will move straight away from the dog. If the stock goes to the right, it's because the dog is left of the balance point or vice versa. As the dog teeters more to one side, or beyond the balance point, the stock turns in the other direction.

When your horse enters a cow's flight zone and encourages her to move, she chooses a direction in which to go. If you block that path, she usually turns her head in the direction of another route. That's your signal that the

The red horned cow was traveling parallel with me, opposite the other cows' direction until I passed her balance point, and she turned in spite of the distance and the cattle between us.

Again, I ride to avoid the flight zone of cattle I don't need to disturb and to minimize the effect my horse and I have on cattle around those that must be moved.

balance point is shifting. If your horse is not within her flight zone, but is teetering on her balance point, she stops, because you're not motivating her to move.

A seasoned cow horse watches a cow's ears and head for signs of a balance-point shift.

Imagine you're in a stream with a beach ball, the current pulling the ball toward you. If you apply finger pressure in the center of the ball to stop it from coming toward you, the ball doesn't go right or left—it's balanced on your finger. This is the balance point. If you move your finger to the right of the balancing point, the stream, which represents the herd in this analogy, pulls the ball (cow) to the left or vice versa.

We can use these principles to our advantage to handle cattle. We use the cow's flight zone to get her to move, and we use the balance point to direct her. Keep in mind, however, that there is a point when the cow feels she's being taken advantage of, and her self-preservation instinct kicks in. She might get on the fight or go through a fence, or do something else we don't want her to do. That's just because we got greedy and tried to get too much done. If we violate the cow enough by staying in her flight zone too much or too long, get her too exhausted, etc., at some point it comes back to work against us.

Nothing Stays the Same

The flight zone is where you engage the cow's sense of self-preservation, but you can't just draw an imaginary circle around the cow and say, "I'm 10 feet away, so I must be in her flight zone." It's that sense of self-preservation—not a specific distance—that comes into play.

Let's say a group of cattle are at rest in a pasture, and you are on foot wandering around through that field with no regard for the cattle. If you are quiet and casual, you probably can get fairly close to the herd without engaging their sense of self-preservation because you aren't threatening them in any way.

But if you enter that same pasture and start raising a fuss and yelling, even if you're standing much farther away, those cattle get up and leave. Their response is as much about your approach as your physical distance from the cattle.

Take the Time

Although it's becoming a lost art, taking the time to train cattle to handle quietly and horses to handle effectively is well worth the effort. Taking the time to school the cows, horses, and riders actually can be a shortcut in the long run.

Like everything in life, there is a balance to working with livestock. One day we do everything we possibly can, and it's still not enough. Other times, it seems we barely do anything and it's too much. If we lose our egos and increase our patience, the livestock teach us what we need to do to be the most effective. We need always to look for that balance.

This is why I can often side-pass my horse in a herd and get really close to rope a calf if I'm subtle about swinging my rope. If I approach the herd with my horse moving quickly or aggressively, that calf leaves before I even get close.

"I enjoy working cattle on horseback because it's a mental challenge for me."

It's safe to say that any time your horse's feet move faster than the cow's feet, you probably alarm the cow.

Practical Applications

When you work cattle outside a corral, you generally have to pay more attention to how fast you get the cow going by noticing her flight zone and how deeply you get into it. You can cut a lot of tough cattle in an open rodear by reading the flight zone and not getting the cow out of a walk, just by blocking a cow when she goes a direction you don't want her to go. You keep blocking her whenever she starts to go any direction you don't want her to go, and soon she looks out of the herd and walks out because she's tired of running into you. This is what I mean by "training" cattle.

I've ridden a green horse to cut a lot of cattle that already have run by people on more experienced horses because I could position myself on the cow's flight zone and balance point accordingly, instead of trying to force things to happen. When things go too fast, it's easy to just get out of her flight zone. I don't have to have a cow on the edge of the herd to work her. I can condition her with 20 cows in the herd between us because she's going to be looking over those other cows and checking to see where I am. When I was a kid, there were plenty of people who could work tough cattle on horses that weren't that handy because the people knew where to be to position the cattle.

Position is especially important when cutting out pairs. If they're too stirred up, they split up and go hide. You have to be careful how much you get in the cow's flight zone so that she keeps her calf with her and goes in the direction you want her to move. Cattle learn really fast.

Unfortunately, the more people work cattle in a corral, the more they tend to lose regard for the flight zone or balance point. I hate to see this because it diminishes the art of working cattle.

I realize not everyone likes the challenge of wondering what a cow or horse is thinking, but I enjoy working cattle on horseback because it's a mental challenge for me. If I can't do something with a cow or horse, I never want to "get a bigger stick" and make things happen. I have to try to outsmart that cow or horse, and use their natural instincts to my advantage.

Cattle Learn, Too

As we noted earlier, both cattle and horses work off principles of pressure and relief. This was very obvious in a Barzona steer I had for about three years. The kids raised him on a bottle and even broke him to ride, and he turned out to be a very useful training tool with my horses.

If I needed to practice roping, I could jump toward him from about 20 or 30 feet away, and he took off running. I would rope him by the horns and as soon as the rope came tight, I didn't even have to dally. He immediately stopped, and I could walk my horse up beside him and take off the rope. He'd stand right there as I walked away and rebuilt my loop. Then when I was about 20 to 30 feet away, I'd turn around and come toward him again—and he'd take off. He couldn't have been a better training tool; it was just like he'd read the book.

I could drive him into a fence corner, and the closer I got to him, the faster he'd try to run by me. The farther away I stayed, the slower he moved; he might just trot. I could regulate his speed, which helped with training green horses. It might seem as though the steer was exceptionally smart, but he didn't handle anything differently from an untrained cow in the pasture; he was just more predictable and more refined. He did these things for the same reason any cow or steer does—he moved away from pressure and sought comfort and relief. He was just more experienced at doing this than most.

It all comes back to relief. The way I trained this steer was by never working him until he was tired. As soon as he started to tire, we quit. It was like a game for him.

Everybody has an idea how to drive a cow and maybe block or stop one, but when one turns too fast…

... I need to slow the cow's turn by getting farther out of her flight zone. Note the cow trotting. Turning away creates less of a threat to the cow.

After the turn, I can still maintain my position on the cow's balance point and distance from her flight zone. Note the cow walking.

"You must feel and see where the balance point is according to the cow's expression and her movements — it's not a physical point on the animal."

7

Theory and Reality: Cow Horses and Controlling Cattle

Pointing my horse's head at a cow influences the cow differently than approaching her with my stirrup or my horse's hindquarters leading.

Once you understand the physical and psychological effects your horse has on a cow, you can use the knowledge and experience the horse already has to your advantage.

There are several parallels in what people refer to as "herd instincts" and "cow sense." "Herd instinct" refers to how your horse relates to other horses, while "cow sense" describes how he relates to cattle. If you can identify the two and understand the connection, then you can allow your horse to make the association, tapping into his previous experiences in controlling the speed, direction and position of another animal, as well as himself.

"Think of the flight zone as determining position on a latitudinal line, and balance point as determining position on a longitudinal line. The point where the lines intersect determines the position your horse needs to be to stop the cow."

Using principles he already understands, and presenting them in a way so that he can find relief, minimizes his confusion and fear, and can develop your horse's cow sense, motivation and confidence in a very short time. However, it's your responsibility as the rider to help your horse connect these herd principles to the cow.

Theories in Action

In the last chapter, we introduced the basics of reading cattle while using the flight-zone and balance-point theories. As you put these principles to use, you discover how they influence a cow's movement and can be used to enhance your horse's handiness.

When a horse approaches a cow, she engages her self-preservation instincts and moves away from the horse. If the horse remains in her flight zone, she speeds up. If the horse retreats out of the flight zone, the cow starts to slow down and eventually stops when she no longer perceives the horse as a threat.

Let's consider the scenario when a cow is lying down in the pasture and you ride toward her on horseback. As you approach her flight zone, she starts to feel a little threatened and rises to her feet. If you inch deeper into the zone, she turns and walks away from you. If you continue to invade her space, or move quickly and aggressively, she takes off trotting or running. The deeper you go into the cow's flight zone, the more intensely she responds. Remember: A cow's breed and how she's been handled (or not handled) influence her flight zone.

Spooky cattle and/or those that haven't been worked have larger flight zones than those with calm dispositions and that have been handled. Also, there are times when the cow comes toward you to get back to the other cattle. This can put you inside the flight zone, even if you're stationary or retreating. Regardless of the circumstances that put you there, the deeper you're in the flight zone, the more flight or fight you create. The farther out of the zone you are, the less reaction you cause. Cattle are like horses, especially when they're younger, and their curiosity can be a strong factor. By staying outside or to the edge of the flight zone, you can engage the cow's curiosity and put it to work for you.

Just as you do when working a horse from the ground, if you can draw that cow to look toward you, you can make her step her hindquarters away from you. You also can simply change your position and draw her further out and make her want to come toward you with her front end and move her shoulders toward you.

To turn a cow away, you need to be in position on or in front of her balance point. The cow can shift the balance point with one step, and your horse might need to take 10 steps to get to the new balance point.

Sometimes, the easiest way to move a cow softly is to move your horse's hindquarters toward her. Then, your horse's forward motion moves out of her flight zone.

Cattle are highly sensitive to body language in another animal. Something as subtle as the direction your horse's nose is pointed can influence a cow's movement, especially in a closely confined area. For example, if I'm working in an alley and my horse's nose is

Instead of approaching the cow with my horse's head, I can approach the cow with my stirrup leading...

......or even with my horse's tail leading. Not only does the cow move away more softly when I lead with the stirrup or tail, but as I ride forward, I'm moving out of the flight zone.

tipped into the alley, the cattle perceive it as a threatening act of dominance and don't pass by the horse. But if I tip my horse's nose to the outside, the horse is seen as submissive and the cattle proceed down the alley.

Making the Connection

To effectively work cattle out of a herd, you and your horse need to learn to read a cow's body language and anticipate her flight zone and balance-point shifts so you can influence her movement in the direction you want her to go and at the desired speed. This is also a good exercise to enhance your horse's focus, reaction time and versatility.

My training philosophy involves teaching a horse discipline and motivating him through self-induced pressure and relief.

Think of the flight zone as determining position on a latitudinal line, and balance point as determining position on a longitudinal line. The point where the lines intersect determines the position your horse needs to be to stop the cow. So how does this create pressure and relief on the horse?

When the cow is on offense and your horse is on defense, the cow stands still when your horse is in the correct position—that is, out of the cow's flight zone and on the balance point. When the cow is relaxed and quiet, you want to allow your horse to relax and stand quietly. Don't distract him by pulling on the reins, kicking him or spurring. This lack of pressure is a major motivator for him to stay in position with that cow. When your horse is focused on the cow, staying out of her flight zone and on her balance point, offer your horse this relief, and he comes to associate the cow with comfort and security.

Myths and Realities

There are some faulty assumptions—call them myths, if you like—that people have about working with cattle. For example,

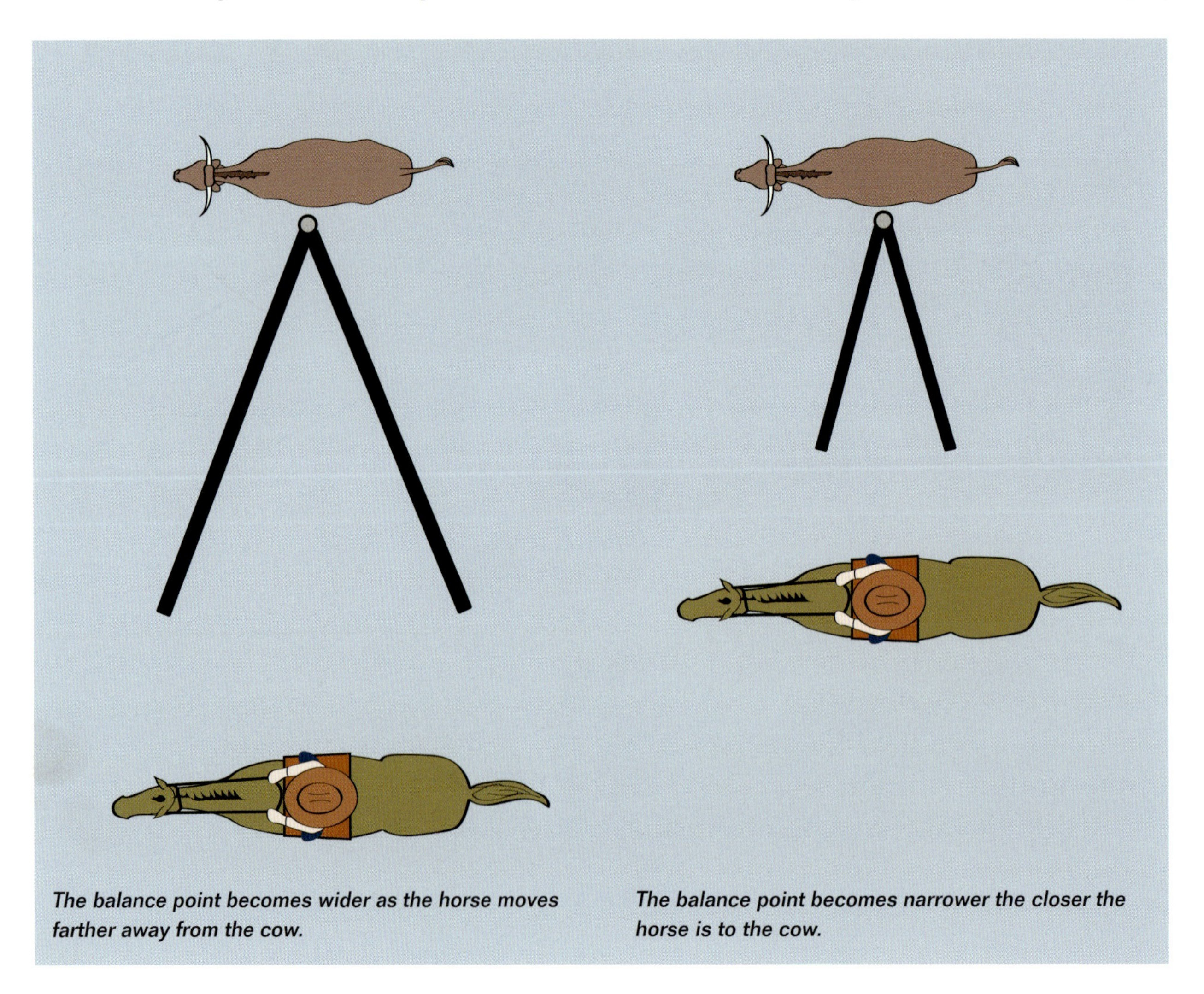

The balance point becomes wider as the horse moves farther away from the cow.

The balance point becomes narrower the closer the horse is to the cow.

As I approach the balance point, the horse's hindquarters are farther away, inviting the cow toward my horse.

As the cow turns, her hindquarters step out and her front steps into the turn.

My horse mirrors the cow's steps and follows her through the turn. The more the horse or cow steps out, the more the turn slows.

cutters do many things that people just assume reflect how you work cattle, and ropers do things that people assume are part of how you're supposed to rope cattle. Coming from a different background, it's easy to take an objective view and realize some of these things are done because of style or to appeal to a judge in a competition. There's more showmanship and more consideration of time because these riders work against the clock.

In a cutting competition you can stay behind the balance point of the cow because you never have to turn the cow. You can follow the cow across the pen until the cow turns off the turn-back or corner men. I think it's fair to say a big percentage of the turns in the cutting pen are initiated by the turn-back riders and not by the cutting horse, but this doesn't happen on a ranch.

On the ranch, when you're working a cow out of the herd, you don't have four other riders help you work one cow; there is one person working one cow. Once you get control of the cow, someone might assist you in driving her out of the herd so you can go get another one, but it's totally different from working cattle in the cutting pen.

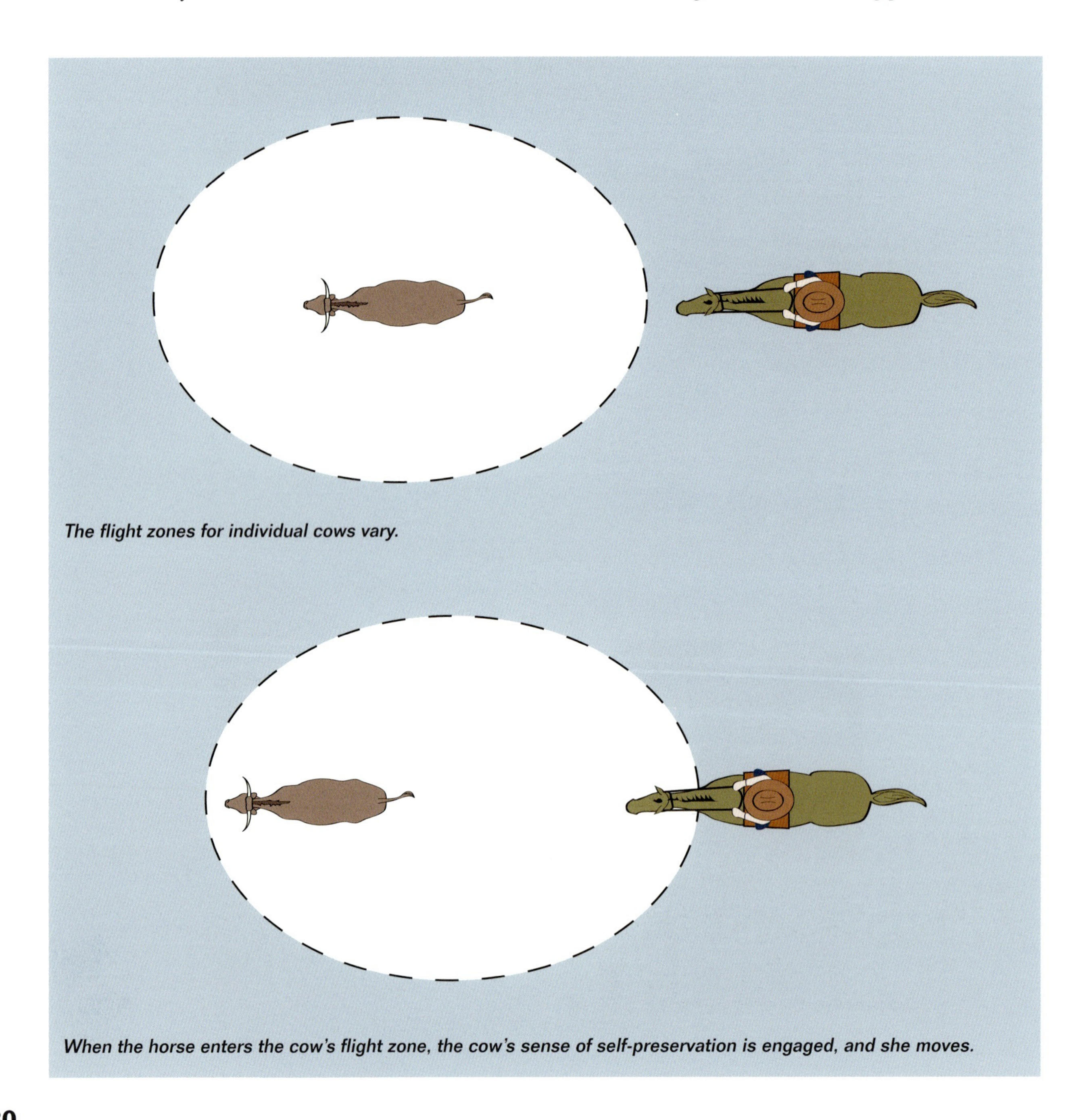

The flight zones for individual cows vary.

When the horse enters the cow's flight zone, the cow's sense of self-preservation is engaged, and she moves.

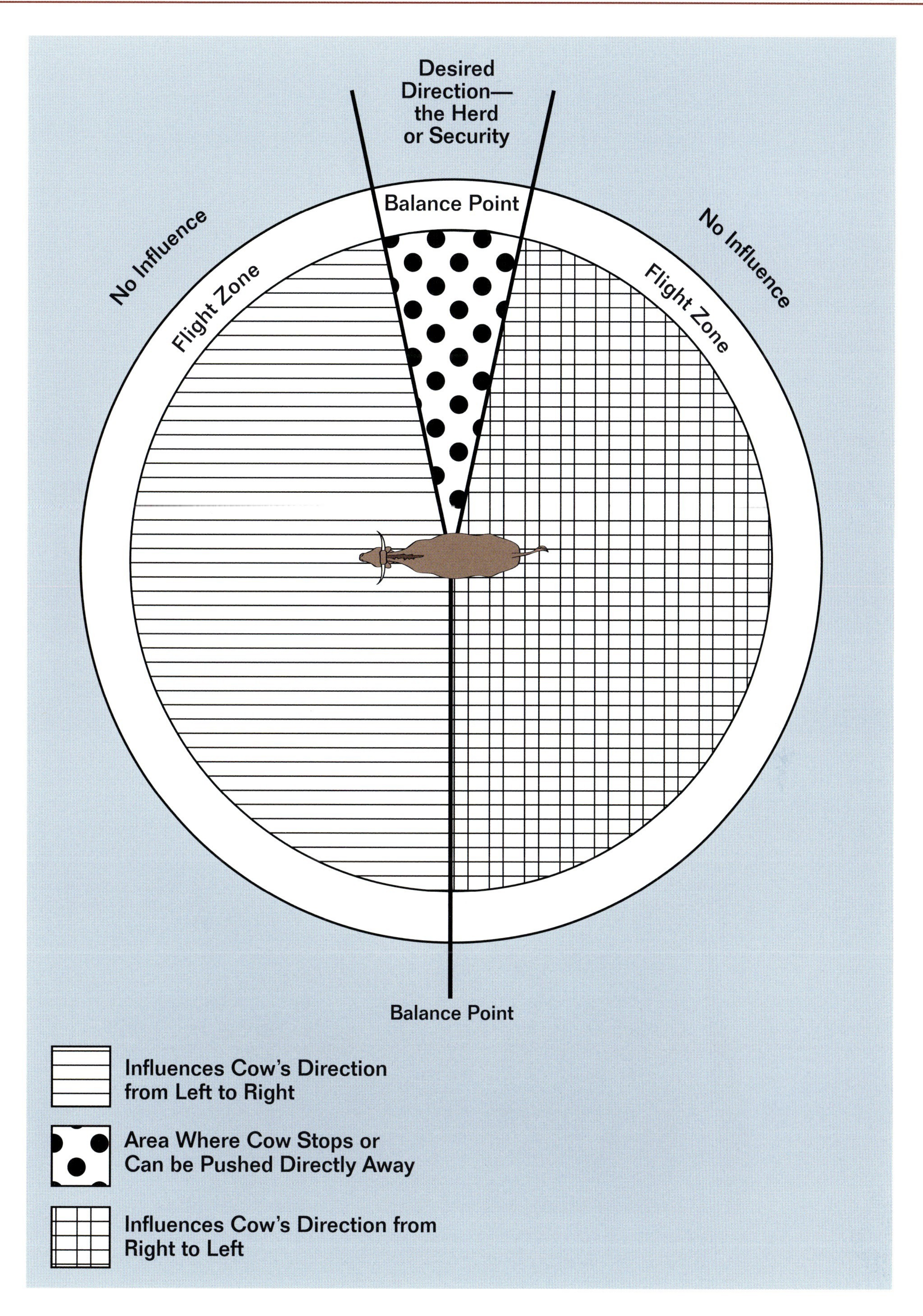
Desired Direction—the Herd or Security
Balance Point
No Influence
Flight Zone
No Influence
Flight Zone
Balance Point
Influences Cow's Direction from Left to Right
Area Where Cow Stops or Can be Pushed Directly Away
Influences Cow's Direction from Right to Left

Two things you hear a lot in cutting are the terms "inside" and "outside" the cow. In a defensive cutting situation, many people want to train the horse to "mirror" the cow and maintain a consistent position with the horse's head on the cow's neck or the stirrup at the cow's flank. When the rider picks a point on the cow, if the horse gets ahead of that point, that is referred to as being outside the cow. As long as the horse stays behind that point, he's said to be inside the cow, and this is where a cutter wants the horse to be.

Sometimes people define outside the cow as being ahead of what I refer to as the balance point, and inside the cow as being behind the balance point. But there also is a point where

With the horse's head being the closest point to the cattle, they are more intimidated and stay farther away from me.

By positioning my horse so the cattle pass by close to the horse's tail or my stirrup, they pass close to me and are more relaxed.

the horse and rider are not ahead or behind, but actually have the cow balanced. That's the point I like my horse to recognize and operate from. So there are basically three positions a horse and rider can take on a cow:

- ahead of the balance point;
- behind the balance point;
- on the balance point.

One myth is that the balance point is a certain, specific point on the cow, but it's much more than that. When a person can't read where the balance point is, he often invents one by making an imaginary point on the cow. For example, he says the cow's shoulder is her balance point, or it's where his stirrup is even with her flank. The problem with this is that the balance point is always changing. You must feel and see where the balance point is according to the cow's expression and her movements—it's not a physical point on the animal.

If you drive a cow away from the herd and down a fence, the balance point is the spot where, if you put your finger there, you could push the cow away. That spot might be on her hip or flank in this situation. But if she's coming down the fence toward you, heading back to the herd, that balance point likely is farther ahead on her body, at her shoulder or even her nose.

Another myth is that the balance point is a specific size. In reality, the farther you are from the cow, out of her flight zone, the wider the balance-point area becomes.

"A horse is born speaking the language of flight zone and balance point because he has been reading other horses his entire life."

The flight zone also changes and the only way to tell when it does is to read the cow. A cow tells you where the flight zone is, the same as she does with the balance point, and either can change with every step. If

Approaching cattle broadside to press, or push, on them leaves less distance for them to see around me. I can affect a broader area and not have to move so much.

Cow Control Exercises

These simple exercises greatly enhance your understanding of how the horse's position influences the flight zone and balance point and, therefore, the movement and speed of the cow. Both exercises can help you discover how much control you really have with a cow, and how to improve that control.

Remember, it's never about "chasing" the cow. Instead, it's all about getting in or out of her flight zone and knowing where the balance point is. Once you understand how to mentally prepare the cow to do what you want, you can reduce stress on everyone involved, including cattle and horses, not to mention humans. You don't need to get a group of people or a hot shot to push and control cattle if you can mentally direct them to physically move where you want them to go.

Exercise #1: Place two rocks or other "target" objects at two locations in a field. Then use your horse to direct a single cow or steer to move toward one rock, and turn the cow away from you, for example, moving to the left, around the rock. Now, position your horse to move toward and circle the one rock. You then turn to the right around the other rock. Basically you want to "figure-eight" the cow around the two rocks. Then turn the cow to "figure-eight" around the rocks in the opposite direction. Do this by staying on the same side of the rocks, or between your cow and the herd, so you are turning the cow into and away from your left and right.

Exercise #2: Take a piece of baling twine and lay it on the ground in the pasture to make a circle. Use your horse to move a single cow or steer across the pasture and get that animal to stop with just one foot—it doesn't necessarily matter which one—inside the circle of twine. Start with a larger circle and work down to a circle half the size of the cow's step, about 18 inches or so.

After you have accomplished this, you can increase the challenge by trying to get the cow to put a specific foot in that circle. You learn so much about positioning your horse and how you and your horse affect a cow's flight zone. If you can't get the cow to stand somewhere near that circle, you're not going to have the tools to get her to go through a gate if she doesn't want to go.

you cause the cow to speed up, it's because you're moving deeper in her flight zone. If you cause her to change direction, it's because your position is changing in relation to her balance point.

When you drive a cow along a fence line, if you're behind the balance point, she continues moving forward. When you get ahead of the balance point, she turns on the fence. If you are on the balance point, the cow stops.

The cow clearly tells you where you are in relation to the balance point.

Other factors also play a role. For example, the balance point is different when you work a cow against a fence than it is in an open area. Also, if the cow comes toward you aggressively, the balance point differs, as compared with when she's farther away from you. If you're deeper into the flight zone, her reaction is different from when you're outside the flight zone. You never deal with only one factor; all the factors have to be calculated. It takes experience working with cattle without the aid of other people or fences to truly understand the balance point and flight zone.

Think of a reined cow horse taking a cow down the fence. If the horse is shoulder to shoulder with this cow, the balance point changes when the horse moves only an inch forward or an inch back. But if that horse is several feet away from the cow, the balance point is larger, and if the horse is 50 feet away, the balance point could be as wide as 20 feet. The farther away the horse is from the cow, the wider the balance point becomes. The balance point actually expands in a "V" shape as the horse and rider move away from the cow.

It's a myth that a horse can learn to read cattle by working any type of mechanical device. It's common to use a "flag" or mechanical cow, or some type of controlled device. The only problem with this is that an inanimate object has no flight zone or balance point so the horse cannot "read" it. The rider must guess where the flight zone is going to be on the imitation cow, or flag, and since the flag can't respond to the horse's actions, this can be confusing for him. He can't develop any sense of flight zone or balance point when working a device. Because a mechanical device has no "feel," the horse must operate off repetition designated by the rider.

Some people might like the mechanical control they have with a flag or similar device, but I think a real cow is better for the horse; I don't know a person alive who can read a cow any better than a horse can. A horse is born speaking that language of flight zone and balance point because he has been reading other horses his entire life.

It's much like a doctor who trained in a foreign country and comes to America. He's still a doctor, and once he understands English, he simply takes all that medical knowledge he already has and translates the language. The horse takes the knowledge and experience he has reading horses and transfers it to reading cattle.

In addition, with a real cow you can dial up how much activity you want by getting deeper in the flight zone, or by having your turn-back riders get deeper in the flight zone, if you want to speed up the cow.

In my opinion, a cow horse that learns to work a cow by reading the flight zone and balance point, and how to handle himself off those two components, is prepared to handle any situation with a cow.

That said, I don't necessarily think it's detrimental to work a horse on a mechanical device. I would compare the artificial cow devices to dry work, and I think it's fair to say that we need a certain amount of dry work to help our stops and turns with the cow. But once the horse learns that the objective is to stay out of the flight zone and on the balance point, then the horse working cattle takes very little direction from the rider.

I had one of my students get on a green horse that I'd ridden on cattle only about 30 to 45 times. The horse was nowhere near finished, but he was pretty keen on reading a cow. This particular student was a good rider and actually had been working with a cutting trainer, but he quickly realized that the green horse was way ahead of him in reading the cow. The horse controlled the cow on his own, without waiting for direction from the rider, and the only way the horse learned to do this was by working with live cattle.

Not Just for Cow Horses

These principles aren't just for cutting or working cow horses; they can be applied to roping, team penning and everyday ranch work. I've even used them to train racehorses, reiners and jumpers. It's important, however, to stay out of your horse's way and encourage him to side-pass, back, etc. You also must know how to maneuver your horse's front end and hindquarters. The less you interfere with your horse, the more he can do the things he already knows how to do, or what comes naturally to him.

After the cow is blocked from going right or left, she stops. This is a good opportunity for my horse to pause and realize his success. Notice the cow's poll is tipped to her right, preparing for a right turn. I don't need to interfere with her thought process by distracting her. As she turns, I will follow her softly away from the herd.

"One of the first lessons a young foal or calf learns is to follow its mother or the herd for security. As the animal ages, it begins to play aggressively with herd mates, with one animal expressing dominant body language until the other submits, thus beginning each animal's offensive and defensive experiences."

8

Offense and Defense

Working cattle, whether on a ranch or in an arena, is akin to a competitive team sport. You and your horse are on one team, and the cow is your opponent. Each team assumes offensive or defensive roles.

Think of the cow as the opposing player. The cow's action dictates your horse's reaction, and vice versa. For example, if the cow turns into you, your horse has to give more ground and work more like a cutting horse. When the cow turns away from you, it's more of an offensive maneuver, and the horse can move more toward the cow. This can be confusing to you if you aren't

Here, my hands are soft, supporting the horse's direction. The horse's feet are in sync and his body is arced as he mirrors the cow.

thinking about how to position the horse on the cow to make that cow do what you want.

Your horse's position can influence the outcome of a play. An experienced cow horse can read a cow's body language, anticipate her next move and outthink his opponent, assuming either an offensive or defensive role. He adjusts his position so that the play works to his advantage, not the cow's. If properly prepared like a professional athlete, your horse can learn to read and anticipate a cow's next move as well as or better than most people can because the horse and cow speak the same language.

"With consistently positive experiences, your cow horse learns to dominate the cow and finds that encouraging. He develops an interest in controlling the cow, reading her body language, anticipating her actions, and gauging his position."

Raised in a herd environment, horses and cattle develop self-preservation instincts from a young age. One of the first lessons a young foal or calf learns is to follow its mother or the herd for security. As the animal ages, it begins to play aggressively with herd mates, with one animal expressing dominant body language until the other submits, thus beginning each animal's offensive and defensive experiences. For example, when a horse lowers his head, pins his ears and makes a dominant movement toward a cow, the cow understands that the horse is threatening her.

To establish clear communication for your horse to handle the cow, you must:

- present the offensive and defensive plays in a way your horse understands and can relate to past experiences;
- build his confidence to call the right play;
- and avoid moving the reins, spurring him or needlessly shifting your weight in the saddle, which distracts your horse's attention from the cow.

On the Offense

Any time your horse is being aggressive and moves toward the cow, using the flight zone to dictate her speed and the balance point to keep her moving in the direction you want, he moves offensively. Common examples of offensive maneuvers include fence work in ranch-horse versatility and reined cow horse competition, or tracking a cow to rope her.

If the cow challenges your horse by trying to return to the herd, the horse should hold or give ground, resisting the urge to move forward. This is assuming a defensive position, blocking the cow enough times that she submits to his control, which we get to later in this chapter.

Teaching a young horse to track a cow is the most elementary form of offense. Before you begin the mechanics of teaching a horse to track, think about how tracking another animal relates to something already familiar to a horse, such as his relationship with his dam or herd mates. As a foal, your horse instinctively sought security from his mother, and played offense and defense with his herd mates. When he felt threatened, he returned to his mother's side. You can tap into his past experiences, using pressure and relief, to teach your horse to lock onto a cow just as he did when rating his mother.

For the first lessons, select a cow that moves away as your horse approaches, but isn't too flighty. If you are on a young horse that hasn't had any exposure to cattle, guide the horse toward the cow. Any time he's pointed in the cow's direction, relax your rein and leg pressure, so your horse feels relief. If he turns away from the cow, direct him a little more intently and use leg pressure. Release that pressure when he moves toward the cow.

At this point, you're not asking your horse to track the cow, just to make the connection that the cow offers him the same feelings of comfort and security that his mother did, and that moving away from the cow results in his discomfort. Once your horse makes this association, he's tracking a cow on his own.

The cow's ears are forward as she looks for an escape from us.

As we move into the balance point, my horse is in time with the cow, and her ears come back to a submissive position.

Here, I'm picking the horse's nose up to create some discomfort while we are out of position.

Get Back on Track

After your horse willingly tracks the cow, ask him to trot or slow lope behind her, using her flight zone to keep her moving forward. To encourage your horse to lock on to the cow, use rein and leg pressure to guide him. When your horse's nose is pointed toward the cow and his attention is on her, indicated by him watching and moving with the cow, ride on a loose rein and relax your legs. You want him to feel relief as though you're not on his back. He learns to associate that relief with the cow.

Remember that your horse doesn't have to be right on the cow's tail to focus on that animal. You must learn to recognize the slightest indication that he's thinking about the cow, even if he's 20 feet or farther away from her. If your horse isn't focused on the cow, or decides to move away from

This horse is going past the balance point and is out of sync with the cow. I use a firm pull on the rein with a quick release to bring the horse back to the cow's balance point.

her, allow him to do so for a few steps. Once he commits to making the mistake, apply pressure with your reins, legs and spurs—as though a predator is going to get him—and redirect his attention back on the cow.

Give your horse the benefit of the doubt and see if he finds the solution on his own, rather than being quick to correct him. This promotes confidence and good judgment, while correcting him too soon can intimidate him and cause self-doubt.

Imagine that a rope connects your horse's head to the cow's tail. When your horse leaves the cow, the rope tightens and the horse hits the end of it within a few steps. That simulated pull at the end of the rope offers him relief when he's pointed back toward the cow. Experiencing discomfort whenever he moves away from the cow and relief when he's with the cow motivates your horse to seek the good feeling associated with staying focused on the cow.

When the horse is in sync with the cow and maintaining the balance point, my hands are soft and supportive.

This sounds logical, but it's the opposite of what some people do. Some riders are quiet with their hands and legs until the horse gets close to cattle. Then, the rider gets busy with his hands to keep the horse right with the cow. As a result, the horse associates being close to the cow with discomfort, because there's more pressure from the rider when the horse is with the cow than when he's away from it. Don't use pressure to direct him on the cow. Use pressure only when he commits to leaving the cow.

> **"If the horse feels good, "hunting" the cow can become a game to him, and he starts becoming more aggressive and quicker to get to the cow."**

This approach yields quick results for most horses because you capitalize on a behavior or instinct the horse already understands. He just needs you to connect that behavior or instinct to the cow. The horse soon learns to go toward the cow's tail for comfort and security anytime he feels pressure.

Usually, in just a few rides a young horse goes from being unsure of what he's supposed to do with the cow to being aggressive, pinning his ears and going toward the cow. If the horse feels good, "hunting" the cow can become a game to him, and he starts becoming more aggressive and quicker to get to the cow. When this happens, you don't want to overwork your horse to the point he's too tired or loses interest.

Some horses might be afraid of cattle or just uninterested, but once they learn they can dominate the cow, they pin back their ears and get aggressive. This is a good sign that the horse is getting confident. When your horse shows such confidence, give him a recess and set up the situation later. If you overwork your horse, you discourage him from wanting to play the game. The trick, as with other things you teach your horse, is to determine how much work is enough. You don't want to do too much, but if you don't ask something of him, nothing happens. You need to find the most effective balance, and that's something only your horse can teach you.

A confident cow horse reads, anticipates and outthinks a cow with little or no influence from the rider. If your horse pays attention to the cow, he becomes in sync with her, and his footfall sequence mirrors hers. A seasoned cow horse wants to be with the cow and control her by using her flight zone and balance point. The cow then submits. One obvious indication that the cow is giving in to the horse is when the cow's ears are no longer forward.

Apply the Offensive Plays

When your horse instinctively follows a cow on a loose rein, you can start to increase the

My horse and I are slightly behind the cow's balance point, pushing on the hip and tipping the front end toward us.

speed at which you track the cow and even start circling the cow or turning her along the fence. These all are forms of offense. To increase the cow's speed down the fence, ride deeper into her flight zone, staying slightly behind the balance point and always keeping in mind that the balance point changes.

For example, if the cow focuses on the arena corner coming up, she might change the balance point and stop. However, if she's focused more on the horse than the arena corner, she goes around the corner, or over or through the fence. Cutters and turn-back riders sometimes use this tactic to teach their horses to stay "inside the cow," or behind the balance point, rather than teaching their horses to control the turn the cow makes by going past or outside the balance point.

When the cow shifts the balance point and turns back, remember to keep your weight in your outside stirrup to stay out of your horse's way and maximize his efficiency, as described in the section on "Watch your Weight" in Chapter 4.

Give your horse an opportunity to stop and turn with the cow and get back into position on his own. If he doesn't, apply pressure to encourage him to get right back with the cow after the turn is complete. It is important not to use too much pressure while stopping or turning. The horse needs to be able to think about where and how he is placing his feet. After he gets pointed back toward the cow, pressure him to get right to her. Soon, he locks onto the cow and follows her wherever she goes, without you constantly guiding him. This is the focus and lightness you want in a winning teammate.

Experience develops your horse's confidence and savvy judgment. With consistently positive experiences your cow horse learns to dominate the cow and finds that encouraging. Your horse develops an interest in controlling the cow, reading her body language, anticipating her actions, and gauging his position in relation to both her flight zone and her balance point to control her movement.

As we move on to the balance point, the cow straightens her path of travel to parallel us.

As we move slightly ahead of the balance point, the cow tips to her left and moves away from us. Notice the footfall of the cow and horse are in sync in all three photos. The horse only gains speed and distance in that the length of his stride allows him to maintain the timing with the cow.

Build a Strong Defense

If you and your horse are to have an effective offense, you must develop a strong defense.

In sports, defense is the art of preventing an opponent from entering your territory. The term also holds similar meaning in horsemanship and stockmanship. You apply defensive plays while cutting, boxing or sorting cattle, and when you have to separate a cow from the herd while the other animals remain behind you. Your objective is to keep the cow from coming back to the herd, while the cow's mission is to get past you, back to her security zone.

Compare working cattle to football; each team is either on offense or defense. You and your horse are on one side of the scrimmage line, and the cow is on the other. The herd, or the "goal," is behind you. The cow's movement and position with regard to the herd, gate or other "goal" determines which role your horse must play.

Think of the cow as the player with the ball. If you're in offensive mode, driving a cow forward into a corral, down a fence line or away from the herd for any other reason, and you get out of position, the cow most likely heads back toward the herd, or her goal. To prevent that from happening, your horse must assume a defensive role immediately to block and stop the cow. In some cases, this might require taking only one step backward or turning and taking a step forward to get back on the balance point to redirect the cow.

But, if your horse doesn't go on the defense when needed, you risk losing the cow and might have to run a mile to get control of her again. In a cow horse competition, a weak defense takes you out of contention. On the ranch, this stresses the cattle, spends your horse's energy and costs you valuable time.

The horse is backing away from the cow's flight zone to slow her down. This horse knows it is more work to control the balance point of a cow when she's close than when she's farther away.

Changing Situations

In a defensive situation on any cow, regardless of where I am in the flight zone, my horse must learn that when the balance point changes on that cow, he needs to stop, turn and get back to the new balance point. I might regulate somewhat how deep I want him in the flight zone, but any time the cow changes directions, the horse can assume he needs to change directions. I don't lie to the horse and tell him that this time he needs to stop and back up three steps, but another time he can just turn and go. The horse needs to learn that when the cow changes her direction, then he also has to change his direction.

If the horse is deep in the cow's flight zone, the horse might need to back up a step or two to get farther away from the cow before turning to come back. If that horse already is late and the cow has some distance on him, the horse might just pivot on his inside foot and hurry to come through the turn and maintain some forward motion.

If the cow manages to blow past your horse, you might have a weak defense. The key to a great defense is to position your horse on the cow's balance point, so you don't allow her to travel to the right or left. You also want to regulate the cow's flight zone, so you can regulate her speed to the point your horse can control her. Your horse can control the cow by keeping the flight zone and balance point in sync with each other and with his experience level. Whether you cut or work cattle in an open field, or in a corral, this is the most efficient and stress-free system I've found to work cattle.

Growing up on a ranch, I had to learn to do more with less; it was critical for me to plan ahead and maintain control of my cow. I rode several green horses, and the high desert terrain was brushy and rocky. That's why it was so critical to get a horse in the right position to get the response needed from the cow or cattle. If I was out of position, I had to speed up my horse and ride through environmental obstacles.

I learned quickly to maneuver a cow like a border collie, even if it took five minutes to move around her. I had to position my horse in relation to the cow's flight zone and on her balance point. If the cow started to move too fast as I entered her flight zone, I backed off, adjusted to a new balance point, then re-entered the flight zone with the speed and direction needed to keep her moving correctly.

The horse is focused on the calf's body language and is anticipating the next move.

It was critical to plan ahead and maintain control of the cow because, if I made a mistake and the cow escaped, I might have to spend hours or days riding big country just to get the same opportunity with the cow again. But this gave me time to analyze ways to get things right the first time and to do things more efficiently, with the least stress on the cattle.

Developing Strategies

To effectively work cattle, you first must be able to distinguish if your horse is on offense or defense. Remember that offense is when your horse moves forward, toward the cow, and defense is when he moves backward, or away from the cow. As confusing as that might sound, the difference between a defensive turn and an offensive turn is as simple to the horse as black and white. All you must know is if you're going backward or forward.

"A cow isn't a highly intelligent animal, but it's interesting to see how many experienced horsemen have difficulty outthinking one."

Sometimes a rider, when he concentrates too much on what the cow is doing, has a problem. He thinks, because the cow is going forward, that he and his horse also need to go forward to stay with her. Watching as other people work their horses on cattle might be an easier way to understand this concept. If the horse steps his hind feet forward as he turns, he goes forward, which is an offensive turn. The hindquarters push him forward. If the horse rocks back, moving his hind feet farther underneath him, he pulls with his hindquarters. This is a defensive turn, because the horse is trying to retreat and stay out of a cow's flight zone.

Because you don't see what your own horse does with his hind feet as you work a cow, you need to learn how to feel what he's doing. See the movement of another rider's horse and realize the effect it is having on a cow. Doing so can help you identify from the cow's movement how the horse's hindquarters cause the cow to respond in a particular manner.

Taking things a step further, your horse can pivot on his inside or outside hind foot while turning with a cow, and he learns which foot is the most appropriate for the situation. Keep in mind that your horse's pivot foot is the one that bears the most weight for the

The horse is shifting his weight to pivot over the right or inside foot here. Notice the pivot foot is centered directly underneath the horse.

longest period in a turnaround. Most horses turn faster when they pivot on the inside foot because that creates forward motion. When a horse pivots on his outside foot, it's a backing motion. The horse isn't as fast, but that's okay, because he is moving out of the cow's flight zone and giving ground.

I don't train my horses to use a particular pivot foot when working a cow. Instead, I just teach them to read the cow and stay in position. The cow teaches the horse which foot to use and when. A horse can decide better and more quickly when the rider doesn't interfere with his concentration. I trust a horse to know more about how to be a horse than I do. My job is to teach him his job, motivate him to do it, then get out of his way and let him figure out how to do it more efficiently.

The horse is positioning the left hind or outside pivot foot to be used here. His hip will come to the left and back as the front end comes to the right.

Defense in the Rodear

You see horses initiate defensive plays most frequently in cutting, and while working cattle in the rodear (out in the open), which is discussed in depth in upcoming chapters. In this situation, your goal is to keep the cattle as quiet as possible, especially if you're working pairs. If you draw too much attention, the pair you're working panics and separates, the other cattle around you get excited, and you might separate the other pairs. This is a situation in which being slow, effective and efficient is better than being fast and creating chaos.

A cow isn't a highly intelligent animal, but it's interesting to see how many experienced horsemen have difficulty outthinking one. When you run out of knowledge and resort to using force and strength, you intimidate the cow, which causes her to use survival instincts to run and seek the herd's protection.

If you read the cow's flight zone, however, her speed tells you when to back off her. As long as you can maintain position on the cow's balance point, you can control her direction. The tricky thing with pairs is that the balance points and flight zones, as well as the herd instinct to seek each other or return to the herd, might be different for each animal. Not only does a good cow horse learn to read and control a single cow, but an experienced horse also learns to keep a cow and her calf together while separating them from the herd.

A horse can work several tough cattle slowly if he learns to get out of the flight zone, even if it means retreating deep into the herd, to maintain position on the cow's balance point. If things get out of control, it's generally because you're too deep into the flight zone and too far away from the balance point.

As already discussed, the flight zone determines speed, and the balance point is what determines direction. A horse can learn very quickly how to slow a cow by getting away from her and how to stop her by getting on her balance point. Waiting even a few seconds gives your horse time to settle and learn to get into position so he can relax. At the same time, those few seconds give the cow time to realize that it might be easier to yield to the horse and seek relief somewhere else.

It's critical to let your horse proceed with caution. Too much speed going into a cow's flight zone can cause the cow to resist, and your horse might get out of position. Working slow and steady over the long term allows you to work more cattle on fewer horses, and, as a result, both the horses and cattle work better next time.

Going Against the Grain

Growing up, many of us were told never to turn a horse's tail toward a cow when she changes direction. However, as I gained experience working cattle, I discovered that in some situations the best—or only way—to keep a horse in position on the cow is to turn his tail toward her.

"The cow teaches the horse which foot to use and when. A horse can decide better and more quickly without the rider interfering with his concentration."

First, consider the reason you turn your horse toward the cow or away from her. Obviously, if you're teaching a young horse how to watch a cow, you don't want to encourage him to look away from her. But, if you're trying to maintain position on a cow and turning tail enables you to work faster and more efficiently, then do it.

When a cow comes at you with a lot of determination and confidence, and your horse is at an angle from which you have to stop and make a three-quarter turn, while the cow has to turn only one-quarter to maintain her speed, it's very difficult for you to stay in position to prevent her from continuing on her path. If you get too close before she changes direction, she can speed up while you stop and turn.

Before you can catch up with her and get back into position, she acknowledges the relief as she gains distance from you. As you approach her flight zone again, she looks for you to get into position so she can get relief again. She's being trained to get away from you. If your horse has more endurance than the cow, and she doesn't run through a fence, you might eventually get her stopped when she's exhausted.

Even if you get turned around fairly close to her, you're behind the balance point and in the flight zone, which means you're speeding her up in the wrong direction. Chances are you're hurrying your horse too much and distracting any thought he might have of the cow. If your horse isn't focused on the cow, you're just practicing dry work.

In that case, when the cow commits to the turn, simply turn away from her, maintaining position on her balance point. You also maintain position consistent to her flight zone, and because you don't increase the pressure

As the cow turns to trot away, I turn away from her, and she slows to a walk. Now my horse can go right back into position on the cow. Doing this can help keep the horse and cattle quiet until we have a better opportunity to handle more speed appropriately.

by turning into her, she doesn't speed up. Because you stay in sync with her, you can regulate a more consistent amount of pressure on her, and she doesn't get away from you.

Another example of when it's beneficial to turn your horse's tail toward the cow is when your horse is handicapped by slick ground or rough terrain and turning toward the cow would get him out of position. When your horse becomes handy enough, and the cow slows down and softens up enough, you can start showing your horse the cow by turning toward her.

Sometimes turning tail is actually the best way to get a job done working cattle while giving your horse a quiet learning experience.

As the horse gains experience, he learns to handle tougher cattle head to head. This horse is confidently backing away, maintaining the balance point and giving himself room to turn as the cow commits to the turn.

"There's no part of the "A" pen that most people with a lot of experience working cattle haven't used at some point."

9

"A" Pen: Construction and Theory

The principles behind the "A" pen for working cattle have been around forever. Plenty of times a person drives a cow into a corner toward a gate and the cow refuses to go through it. When she refuses, she comes back to put pressure on the horse and rider. What happens next depends on how the horse responds.

If the horse quickly gets out of the cow's flight zone and gets on the balance point, he can slow her down and actually stop her. His correct response in reading the cow allows him to regain control—and not have to work as hard.

The herd sitting behind the back fence helps draw the cow you're working toward your horse.

This is precisely the situation we try to set up with the A pen, and the A shape makes this easy to do. We literally use the pen's design to set up a situation so that a cow trains the horse. When the pen is used correctly, the horse learns to focus more on the cow, rather than depend on the rider to tell him what to do.

I've used the A pen with many 2-year-olds that have had less than a week's riding on them. The horse learns to stop, turn and move out with the cow much faster than he learns to stop, turn and move out with the rider. This is because the horse learns to rate the cow, based on his previous herd experience with other horses. I've actually had horses with very few rides do little sliding stops, roll back over their hocks and take off after a cow in the A pen when the horses would be weeks away from doing these maneuvers if we were doing them "dry" without a cow. The horse instinctively can understand how to read the flight zone and balance point on a cow a lot easier than the leg and hand cues we try to give to get him to do the same things.

Construction Details

When you're ready to construct an A pen, the crucial thing is the narrow end, which is at the top of the A. You're somewhat limited in size on this end if you want the pen to be as effective as possible. Dimensions can vary, but I've found the narrow end needs to be about 15 to 25 feet wide. That's small enough that if you're up against the cow, she doesn't get relief, but still large enough that the horse can complete a turn, move forward, stop and make another turn.

If you make the top of the A narrower than 15 feet, your horse doesn't have room to turn around. Keep in mind that a horse is about 7 or 8 feet long. If he blocks the cow on the

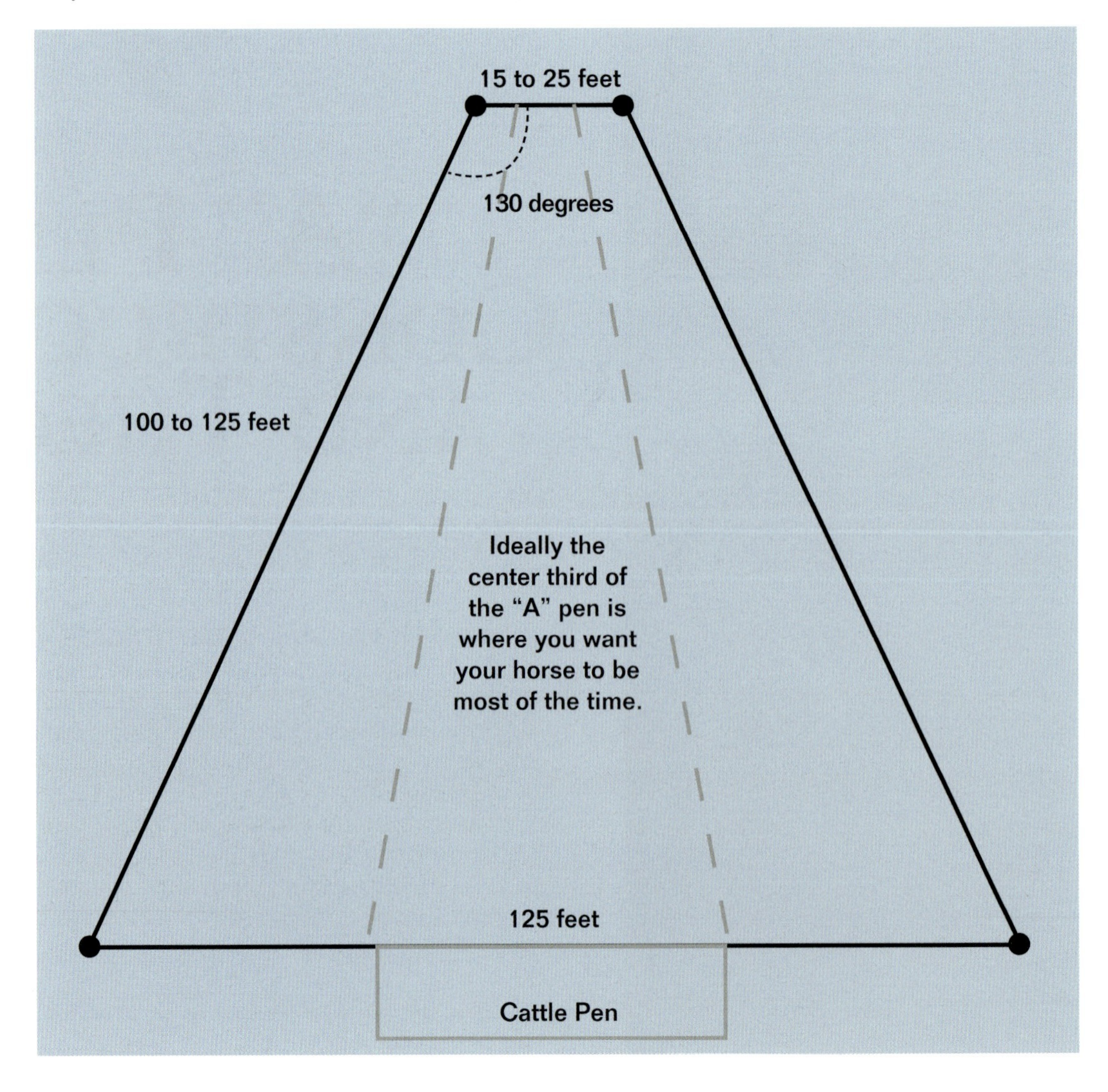

fence on one side, by the time he turns around in an area less than 15 feet wide, he's against the fence on the other side. He simply doesn't have room to get away from the cow very well.

If you make the narrow end wider than 25 feet, this gives the cow too much relief after the turn, and you just end up chasing her up and down the fence. The narrower the end of the A, the more pressure put on the cow. The wider that end of the A, the easier it is for the horse to hold the cow there. You look for the happy medium to give the horse a chance to control the cow, but also to keep pressure on the cow.

Making the narrow end 15 to 25 feet long gives you this happy medium. The horse can turn around, and the cow can still be ahead of him. This gives the horse time to fall away from the cow and get out of her flight zone, but still get ahead of her without pushing her down the fence.

The two side fences of the A should be set about 130 to 135 degrees coming off the narrow end. You don't want to build corners set at 90-degree angles like a typical fence, because this allows the cow to run more, gaining more speed and control and putting your horse at a greater disadvantage. The 130-degree angle coming off the narrow end works best. It gives the horse time and room to get out of the flight zone, but the cow doesn't have time and room to get past the horse.

The side fences need to be at least 100 to 125 feet long. They can be longer, but not less than 100 to 125 feet long. Having longer side fences and a wider back fence gives both the cow and horse more relief. A large pen might be better for a young horse although a performance horse might benefit from a smaller pen, say one with side fences 100 to 125 feet long, and a back fence about 125 feet wide.

The cattle holding pen should be situated in the center along the back fence. This way

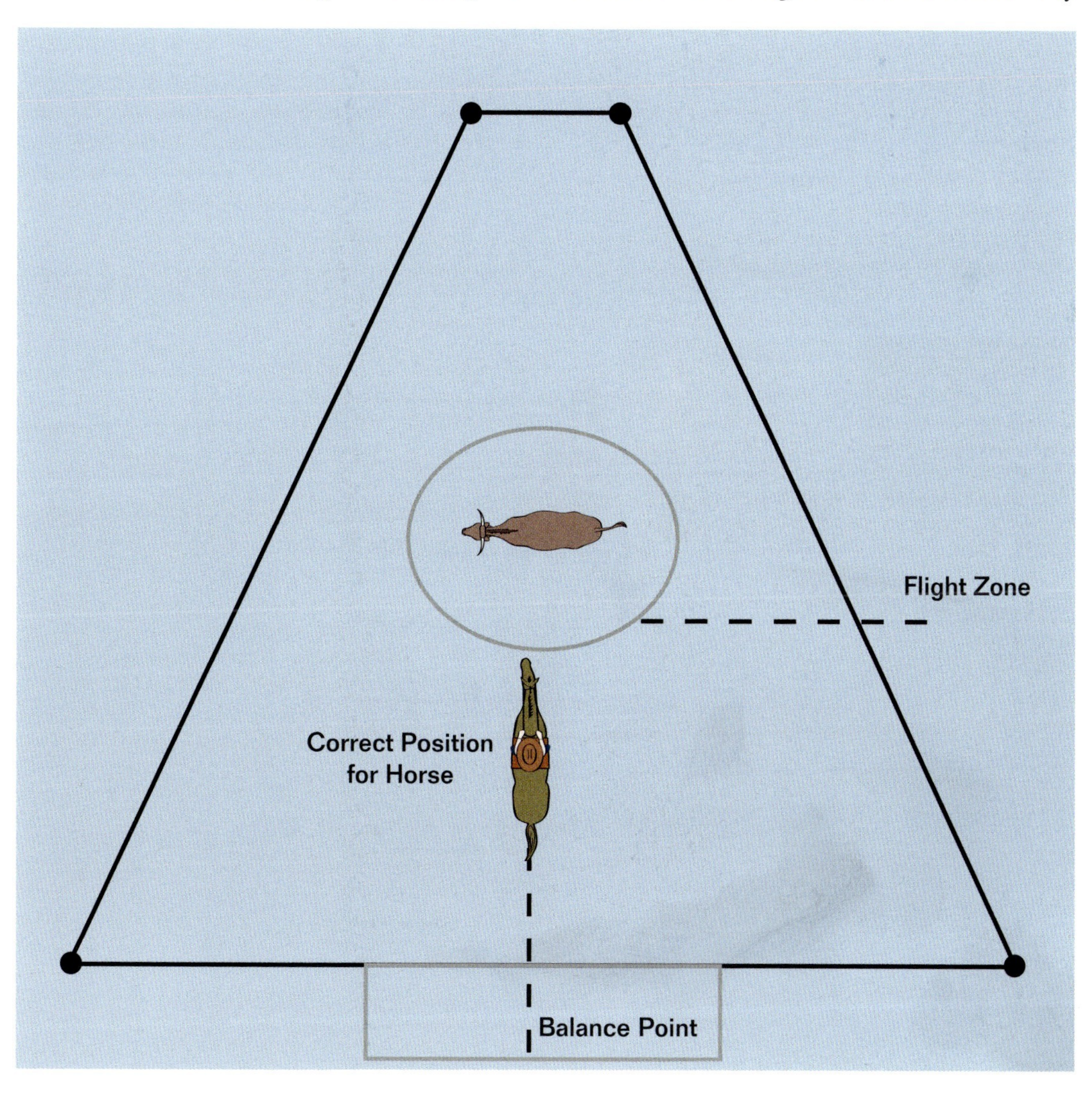

the herd always draws the cow's attention to the center of the pen, which helps your horse to learn to work in the center of the pen. With the cattle behind you, even if you put pressure on a gentle cow, you encourage her to go back to the herd.

"When the pen is used correctly, your horse learns to focus more on the cow, rather than depend on the rider to tell him what to do."

Whether the fence is made of panels or boards, you don't want the spaces wide enough that a cow can stick her head through them. The fence doesn't need to be solid, just watch the spacing.

Theory

Many of the old ranches had big "wings" of fencing extending from their corrals to make it easier to move cattle into the gate opening when riders brought cattle off the range. These fence wings might be 50 to 200 feet long, and they narrowed down to the corral gate. The area in between the wings gave a horse a good opportunity to work from a defensive position if a cow tried to turn back from the gate, and this is actually what is simulated with the A pen.

When you work a cow in the A pen, the farther you get back and away from the corner, the farther you can get out of her flight zone, and the more you can slow her down and even stop her. The farther away you are from the corner, the more you can gain control over the cow.

When your horse stops the cow, you then can let him rest a little, catch his air, and think about what he's just done; the horse learns to seek relief by controlling the cow. He learns to give ground and find a point where he can relax and get his air.

Every time you repeat this, as long as your horse has time to think about getting back for relief, he learns. It's like touching a hot stove; every time he gets too close to the cow and she speeds up, he gets in trouble because he has to work harder and doesn't get relief. The closer he gets to the cow, the harder he has to run and work, so he quickly learns to look for the balance point and to get out of her flight zone so the cow slows down and stops.

The reward for the horse is to relax and evaluate what is taking place.

Teaching the horse to respect these natural principles goes back to what the horse learned as a colt playing in the pasture with other horses. From the cow's standpoint in a working situation, once she learns that no matter how far back she goes or how hard she runs, the horse is on her balance point and blocks her, she gets to the point that you can drive her wherever you want to.

The practical part of this exercise is used on ranches routinely in one form or another. How effectively you can apply the principles and use them to train and improve your horse depends on your horsemanship and stockmanship abilities.

When the horse gets too close, the cow speeds up, we get behind the balance point, and she gets back to the other cattle.

Nothing New

The principles that make the A pen so effective certainly aren't new. They've been around ever since horsemen started working cattle. Actually, there's no part of the A pen that most people who have a lot of experience working cattle haven't used at some point.

I don't like to use the term "invent." Columbus didn't invent America; it was here and it was a discovery to him. I might have discovered something in using the A pen, but I didn't invent it, and I'm not sure somebody else in previous times didn't discover the same thing. Although a lot of people working cattle use different components of the A pen, I've just refined the details and put them all together more than anyone I've run across.

When a horse learns to measure the flight zone and balance point appropriately, he can control the cow in the middle of the pen.

Having said that, my work in the A pen is still an ongoing learning experience. I look back from year to year and realize it's an evolving experience for me. If I'm trying to make a horse into a cow horse, he will spend some time in the A pen.

My use of cattle in the A pen to train horses goes back to 1999 when I was in Australia giving clinics for campdrafting. The nature of that sport is that you cut a cow from the "mob," as they call it, and drive the cow to a gate at the narrow end of the pen. As you approach, the gate is opened so you push the cow through the open gate and out onto a course, where the goal is to direct the cow at speed around markers in an open

The horse can learn to make a good, straight stop...

... rock back over his hindquarters and turn...

area. Visually, it looks a bit like taking the cow through a barrel-racing pattern.

Most of the horses are primed and ready to chase the cow when they get to the narrow end of the pen. But if the horse is thinking too much about running up on the cow, and the cow doesn't see the gate or is thinking too much about getting back to the herd, she tries to turn and run by the horse. If she manages to do this twice, the horse and rider are disqualified.

Studying this, it seemed to me that it was more important not to lose the cow than to get the jump on the cow at the gate. The goal was to teach these horses not to rush the cows in the first part of the competition.

... and then stand quietly...

... until he needs to make the next move.

These horses had a lot of offense in their work, but very little defense. So, in our clinics we left the gate shut and let the horse push the cow against the gate—and the horse would lose her. We went back to basic principles: The horse had to work harder once he got too deep in the cow's flight zone and off the balance point. It didn't take long for the horse to become motivated to keep control of the cow so he didn't have to work so much. Basically, I was just trying to take my stockmanship experience and adapt and apply it to the situation at hand.

Teaching the horse to respect these natural principles goes back to what the horse learned as a colt playing in the pasture with other horses.

In the next six years or so, I did 25 to 30 clinics in Australia, and we've adapted and refined this even more. I really saw the value of how quickly the horse's defense could be improved by using a pen in the shape of an A. I started experimenting until I came up with what I thought were good dimensions, and now the A pen is one of my favorite tools. It's by far the most effective and fastest way to get a horse to read and work a cow defensively that I've ever seen anywhere.

I have started young horses on cattle out of an open rodear, with no fences around, and I use the exact same principles—get out of the cow's flight zone and give the horse rest when he gets on the cow's balance point—that we use in the A pen.

You actually can get a horse working exactly as he would in the A pen when you know what you're doing in the rodear. All the A pen does is help you get more work out of your cattle. But if you're in an open herd and you have 200 heifers to cut out, by the time you work 25 or 30 of them, your horse can be just as keen as if he'd had a lesson in the A pen. You might get only four or five turns on each yearling before he goes out, but by the time you get 25 or 30 head out, your horse has had a good work. It all goes back to stockmanship.

The Importance of Stockmanship

Stockmanship plays a big role when using the "A" pen. If you can't keep pressure on a cow in this tight area, she's too desensitized to be effective for this kind of exercise. If she is too gentle in the first place, she doesn't have a flight zone. If she stands there switching her tail, you've worked her too long and she's just given up.

You must maintain a cow's flight zone; if you work a cow too long, she gets tired and then doesn't have a flight zone. She either quits or gets on the fight, and neither of these scenarios helps you train your horse.

If you get five or 10 quality moves out of a cow in the A pen, consider that enough. Turn her back with the herd and get a new cow. If you let a cow leave a little fresh, she comes back the next day encouraged to keep her feet active because you never worked her long enough to wear her out and make her tired. You get a lot more mileage out of your cattle when you work them this way.

On the other hand, if you work your cattle until they're tired and then quit, they get discouraged, and then they are discouraged every day when you work them. The idea is to develop a confident cow and quit when she feels like she's "won." This was exactly the method I used with the Barzona steer I mentioned in Chapter 6. If you never wear out your cattle, to them working becomes more like a game instead of something they dread.

Most people in the horse-training world have a limited number of cattle to work, which is why they use mechanical cows and devices. But even when you only have a small herd of cattle to use, if you can understand cattle better and learn how to maintain them, you don't have to sacrifice your horse much to get a lot more use out of your cattle. The way I look at it, many people don't want to sacrifice their horse, so they sacrifice the cattle, and then they're trying to work sour cattle. The irony of this: Those people end up sacrificing their horses anyway because the cattle get sour.

There's a balance where you can quit working a cow just a few turns before she quits you. If you can do this, she comes back tomorrow and works better. Then, instead of getting three or four good turns out of her, you might get 10 good works out of her the next day, and she might work all summer for you. It comes down to practicing conscientious stockmanship, not just using the cattle as mere tools.

The interesting thing about cattle is how much they are like horses. If you have an unlimited supply of cattle, it might not be worth working to maintain them. Likewise, if trainers have an unlimited supply of horses, some trainers go through a lot of horses to find a good one.

If you have limited stock—horses or cattle—it might pay to develop the potential of the stock available to you.

As the cow is pushed to the narrow end and sees no other place to go, she tries to come back toward you.

"The horse creates more work for himself by getting out of position, so he quickly learns to get in position on the cow."

10

USING THE "A" PEN TO TRAIN HORSES

I've always had a deep appreciation for a good cow horse, one that has the heart and desire to dominate the cow. It's a very gratifying feeling when my horse is working with me.

While we can't create "cow sense" in a horse, we definitely can build his confidence to work cattle and improve his ability to read them. That's where the "A" pen comes in handy. This isn't about training a horse for cutting or making a finished cow horse, but the A pen is great for building defense, and it develops the horse's desire, confidence and judgment to control a cow in a very short time.

When the horse initiates the turn, he can better measure the cow's balance point than he can when the cow initiates the turn.

Remember: The principles are very simple and the horse learns quickly if we don't complicate things and get in his way. It's our responsibility to present things in a way that make it easy for the horse to learn.

Not Like Cutting

Working in the A pen is one of my favorite exercises to do with students.

Giving a student a cow is like giving a kid a ball; it's fun for him, no matter if he knows how to work the cow or not. I've found that no matter the rider's level—a champion campdrafter or a novice rider—the horse usually picks up on cattle work before the rider does. The hardest thing is getting the rider to turn loose of preconceived judgment calls and let the horse form his own judgment on the cow and make the call.

"The horse learns he can control his own destiny by yielding to the pressure of the cow."

I've had some very experienced riders almost fall off their horses because the horses have never made a move on their own. When a horse stops and sucks back with the cow, it's a foreign feeling to a rider used to being on the offensive with plenty of forward motion.

One problem that surfaces frequently in the A pen is that many people try to implement some of the more traditional cutting school theories, instead of letting a horse read the cattle. When that happens, the results are marginal, at best. I'm not saying that anything is wrong with cutting theories; they're just not meant for the A pen. The principles of cutting and the principles behind the A pen are totally different.

I've found that the less experience a rider has with cutting, the better he or she does working in the A pen. This is because the person isn't trying to "mirror" the cow; he's not worried about being "inside" or "outside" the cow, or stopping straight with her.

Being Effective

When you start working in the A pen, think of the pen as being divided into imaginary thirds lengthwise with one third on each side and one third in the inside, or middle, of the pen. Ideally, you want a horse to maintain control of the cow in the middle of the pen, rather than running all the way to one side and then the other. You want always to give the horse more relief in the middle third of the pen, and more work in the other two-thirds of the pen.

As the horse works in the center of the pen or against the back fence, it's relaxing for him because the cow has more room. In the narrow end of the A, the cow feels more pressure and, in turn, she puts more pressure on the horse as she tries to get past him toward the herd. This generates the same basic effect created by turn-back riders, but the beauty of the A pen is that you don't need other riders to help move the cow.

It takes pressure for relief to be effective and relief for pressure to be effective. This is very much the case when working in the A pen. You offer the horse comfort (relief) when he's in control of the cow. When the cow stops, you let the horse stop, catch his air and process his thoughts. On the contrary, when the horse is not in control of the cow, she's on the move so he has to hustle and work harder (pressure). That becomes a major motivator for the horse to get control of the cow as soon as possible.

Work in the A pen is all about those two topics that we've been covering in depth: flight zone and balance point. These are the two factors that make the A pen so effective.

1. The horse learns to get out of the cow's flight zone.

2. The horse learns where the balance point is on the cow. This stops the cow and gives the horse relief, which is his motivation.

First Lessons

When I introduce a horse to cattle in the A pen for the first time, my goal is to stay between the cow and the herd at the back fence. When the cow first takes off, I direct the horse to leave on a parallel line of travel rather than charging toward the cow. From there, I let the horse dictate how far up the pen we work. Any time he wants to come off the cow, that's fine. If he doesn't want to push on her, that's fine.

When the cow turns off the fence, it's easy for the horse to be late because the horse doesn't initiate the cow's turn as much as the fence does. This can be a no-win situation for the horse to control the cow.

As the horse approaches the cow, her flight zone engages and she moves off. If the horse keeps up with the cow, the cow speeds up, but if the horse leaves her flight zone, the cow can slow down or stop.

When you introduce your horse to cattle, watch which way the cow's ears and nose are directed; this tells you which way the cow is thinking about going. The horse needs to learn to read these signs.

The relationship between flight zone and balance point on a cow is critical. The balance point is the position from which we can influence the cow to change directions. For example, when the cow's flight zone is engaged and her path is blocked, she chooses another route. When her flight zone is not engaged and the horse is on her balance point, the cow stops, which, in turn, allows the horse to stop and rest.

The balance point is always changing, but if your horse can read where the balance point is, you always are able to stop and hold the cow. What determines whether the cow goes right or left is your position on the cow—if you're too far behind or ahead of the balance point. The horse can "tip the balance point" and make the cow change directions sometimes just by moving a matter of inches.

Once the horse learns he can get out of the flight zone to slow down the cow and give himself relief, we have the same effect as if we are turning the horse into a fence to do a rollback. But in this case, the cow's pressure comes toward the horse, pushing him back in the turn. The horse learns he can control his own destiny by yielding to the pressure of the cow.

"...the more the cow gets ahead of him, the harder the horse has to work."

I leave it up to the horse if he wants to move and put pressure on the cow. The horse figures out quickly that the more the cow gets ahead of him, the harder he has to work. He realizes this on his own without me jerking or spurring him. The relief of not working so hard is all it takes to motivate him. He

Whenever the opportunity is there, I let the fence stop the horse. I don't pull him to a stop. Then he can learn to make a nice stop and the turn, if necessary.

starts watching and rating that cow, and learning where he needs to be in relation to her. I want the horse to respond to the pressure the cow puts on him and to the fact that he creates more work for himself by getting out of position. If the horse pays the consequences of having to work harder, he quickly learns to get in position on the cow.

If I have a green horse that is not reading the cow, I get far enough away from the cow

By initiating the turn the horse can rate the turn, make the departure, and be in shape for the next turn better than he can when the cow initiates the turn.

to be out of her flight zone, and then I turn and bring the horse back and forth in front of the cow. I think of this as the "shade-tree effect" because it's like a shade tree casting a shadow on the cow. I put enough pressure on the horse on the ends, turning him back and forth that he wants to seek relief in a comfortable place, which is the middle. My goal is to make that comfortable place on the balance point out of the flight zone in front of the cow.

At first the horse might not associate this comfort, or relief, with the cow, but once he finds that relief, I let him turn and let him face the cow. It usually takes only a few times for the horse to want to turn and face the cow, and from then on he associates the cow with the relief. If I keep offering the horse relief associated with his position on the cow, he relates it to the cow, and that's where I want his focus. Then he starts watching and rating her.

The horse starts using his natural rating instinct on the cow as soon as he realizes the cow offers relief and security—in other words, not having to work hard. In the beginning, I might have to suggest a position to the horse, but rating is natural to him as a herd animal. I can almost see the light switch on in the horse's mind once he picks up on this and realizes that keeping track of the cow is just like keeping track of his mama. It's a sense of security.

I want to build the horse's desire to want to work the cow. He might not always make the right choices, but I want to look for the effort. As long as he makes an effort, I need to direct that effort, not discourage it.

Relief Critical

This is important: The time the horse gets to relax in the A pen should be equal to or greater than the amount of time he spends working. This encourages him to stay in position on the cow.

The relief the horse experiences should relate to his position on the cow—not his position in the pen. Sometimes a cow stops at the same place in the pen every time, but as you keep working different cattle, a horse learns it's not where you are in the pen, but where you are on the cow that leads to relief.

One thing you don't want to do is allow the horse time to relax in a corner of the pen because then he starts thinking of the corner, instead of the cow, as relief. If the horse stops in a corner with the cow, let him stop for a moment, but then move him out of the corner so he doesn't associate that with relief.

Cattle and horses are all individuals. You find some cattle want to work one side of the pen more than others, while some horses are more sluggish turning to one side than the other.

As you work each cow or steer, keep in mind that you want to quit before the animal is ready to quit. Don't get so caught up in the action that you overlook this important detail. After you get a few good turns out of one cow, turn her back in with the herd and select another one to work. Make this a priority so you don't sour your cattle, which is especially crucial if you have only a small group.

The horse's ears indicate he is reading the cow, and I am only supporting his intentions.

Working and Learning

As you work your horse on cattle in the A pen, he learns to make both offensive and defensive moves, but you always want to see a little stronger defense in the horse than offense.

"It usually takes only a few times for the horse to want to turn and face the cow, and from then on he associates the cow with the relief."

As noted, when the horse is in a defensive position, he learns he can relax. When he's in an offensive position, he must work more. Tracking the cow, for example, is an offensive move, but you want to just maintain staying on her balance point. Of course, if you want to speed up things, you can ask the horse to move deeper into the cow's flight zone.

When that cow turns, the horse soon learns he has to hustle to get back in position so he can relax again. This makes him want to get back in position all the quicker. The horse's position on the cow is like the analogy about getting close to a warm stove on a cold day. The closer in position the horse is on the cow—not close to the cow, but in position on the cow—the better the horse feels. You always want the horse to realize the relief he's feeling is related to the cow. When your horse is in position on the cow, he gets relief. When he gets out of position, he has to move and work, and he loses the relief.

I can use my legs to push the horse forward in the turn to work offensively.

Watch your horse. When he licks his lips, this is a sign that he's becoming relaxed and learning to position himself on the cow.

Especially on a young or green horse, don't ask for direction and speed at the same time the horse is turning after a cow. Get your horse going in the right direction first, and then ask for speed. Once the horse figures out this, he starts putting speed into the turns on his own.

You don't want to make "motorbike" turns, which are a common mistake when a rider has a lot of offense in his work. You can solve this by keeping the horse's hindquarters underneath him until you complete the turn and start going forward. The horse can control his turns much better if he's gathered up with his hind end underneath him, rather than stretched out.

A horse needs to pull with his hindquarters in order to turn fast. This means that his hind feet are underneath his center of gravity and more toward or underneath his navel. In order for him to lean forward, the hindquarters quit pulling, which lets his hind feet reposition more toward his tail. Then the turn is powered by the hindquarters pushing forward, creating forward motion.

If we think from the cow's perspective when this happens, instead of a quick turn with the horse pulling back and away from the cow's flight zone, the horse leans forward into the cow's flight zone. Not only does this cause the cow to speed up, but the horse rolls toward the cow while turning, instead of completing the turn and departing away from the cow. This gives the cow a sense of being chased, rather than security, which she feels with the horse retreating as he departs the turn. The cow usually speeds up because the rider asked the horse to accelerate in the turn and relaxed his hands, and didn't hold the horse to keep him from moving forward before the turn was complete. People have a tendency to allow the horse to come out of the turn when

When the horse understands that it's his responsibility to keep track of the balance point …

… I can use the cow to push the horse back in the turn and work defensively.

pointed at the cow, and then try to turn the horse away after he's already moving forward with speed.

Here is a very simple test to check yourself in this situation: If the first forward movement out of the turn is toward the cow, your rein hand still is pulling the horse away. If the turn is completed, the reins are slack, or the cow-side rein is being used to keep your horse from going away from the cow. As you leave the pivoting point of the turn, note if you are reining your horse away from or toward the cow. Reining away means the turn wasn't completed before departure, and reining toward, or neutral, means the turn was completed, or even over-rotated.

Also, when you ride to the fence to head off the cow, stop straight and let the horse settle if the cow has turned away from you to go up the fence. This might give the cow the opportunity to go away and get relief, so you don't need to make the quick turn. Then turn toward the cow after the horse has had a moment or two to settle; this separates the turn from the stop in his mind.

When you must stop your horse, use one hand instead of two. While you certainly want your horse to pay attention to your cues, in the A pen his main focus should be on the cattle at all times. The less fuss you have to make with controlling and directing him, the better.

Dos and Don'ts in the A Pen

Do:

- Have a picture in your mind of the pen divided into thirds. Always give the horse more relief in the middle of the pen, and more work on the other two-thirds of the pen. Try to keep the cow in the middle, rather than running all the way to one side and then the other.
- Leave on at least a parallel line of travel with the cow, or a line that puts more distance from the cow as you travel.
- Let your horse—on his own—rate the cow.
- If you must stop the horse, use one rein. Using both hands with even pressure brings up a horse's head.
- Ride straight to the fence and stop straight and settle, then turn toward the cow after the horse has had a minute to settle. This separates the turn from the stop.
- Put more pressure on the departure than the turn. This is one of the most important things to do.

Don't:

- Distract the horse by pulling on both reins and taking his attention off the cow.
- Spur or kick the horse as he gets closer to the cow.
- Put your horse in a bad spot or get to the fence late so the cow has to make only an eighth of a turn to get around your horse, but the horse has to make a three-quarter turn to head the cow.
- Overbend the horse. He gets out of balance and can't turn as well. Wait for the feet to follow the nose.
- Make "motorbike" turns. Keep your horse's hindquarters underneath him in the turn instead of letting him move forward before completing the turn.
- Work a cow until she's tired.

Do: When the horse honors the cow's flight zone, I can use one rein to bring his nose through the stop to turn with the cow. Using the outside rein has only a negative effect.

Dont: Too much inside rein overbends the horse and interferes with his balance in the turn, which creates a negative effect.

By understanding the principles of controlling a cow with the help of the A pen, a horse can learn to stop and turn, among other things, long before he could ever learn to do those things without a cow.

"Working in the open was the most practical and efficient way to work cattle before fences and corrals were available."

11

Working Cattle in the Rodear

Cattle came to the Great Basin in the late 1800s, but it wasn't until the late 1900s that this area had many boundary fences. It's safe to say that cattle were in the area for a good hundred years with just cowboys controlling their drift, not fences.

From the time I was a kid up until the 1980s, I could ride literally for hundreds of miles without having to open a gate. Sure, there were fences around some privately owned land,

Cattle work out of the rodear if they can see other cattle to go toward, called the pratha. Cattle trained to work in the open can be more relaxed than those worked in pens.

When cattle are worked quietly in the open, it can take less help than when working in corrals.

but there was nothing much in the way of boundary fences. It was extremely expensive to build fences in the Idaho-Oregon-Nevada region. It was cheaper to hire cowboys to herd cattle than it was to build fences.

"The rodear (or rodea) refers to the herd out in the open."

When it was time to sort, brand, or move cattle, everything was done in the rodear. The rodear (or rodea) refers to the herd out in the open. There might or might not be a fence corner to help hold cattle, but it was rare for anyone to have corrals. The very term "rodear" refers to an open setting.

Working in the rodear has been going on ever since vaqueros began working cattle. The practices we used in the Great Basin were used in California for at least 100 years before. This wasn't some novel invention; it was just the most practical and efficient way to work cattle before fences and corrals were available.

All livestock—both cattle and horses—was worked outside. There were horses on the range, and the younger ones might be rounded up and sold every few years. These horses were worked out of the rodear, the same way the cattle were worked.

Because most ranchers in desert areas had a minimal amount of money and materials to build corrals, they built only what they needed to hold livestock right by the ranch. The only corral we had to hold any cattle for shipping was the same corral we used to "rango," or wrangle, our horses. This corral had a loading chute, but was big enough to hold only one truckload of cows.

We usually shipped around 1,000 head of cattle each year, including beef steers and dry or culled cows. When it came time to ship them, we held the whole rodear outside the corral. We sorted off one truckload of animals, put them in the corral and, as they were being loaded, we'd sort out the next truckload. So when the first truck pulled away from the corral, we'd put another group in the corral to be ready for the next truck. However many truckloads we had to load, each group of cattle was worked in an open rodear, one load at a time.

We really didn't have much use for corrals because we were simply used to working livestock in the open. It was much easier to work the cattle outside.

What was done out of necessity in the past can still be done today.

Working out of the rodear is also much less stressful on the cattle when done

properly. Seeing how much easier it is on the cattle and how many more stockmanship skills are required makes this an interesting challenge to me, but the primary fact remains that working out of a rodear is more efficient and easier on the cattle.

If we look at cost versus benefit, it takes less expense and less time to work cattle outside than it takes to build and maintain corrals. In addition to the cost of the structures, we have to take into consideration the loss of the weight on the cattle that results from the stress of working them in a corral. Working in a corral is just easier on people because they don't have to think like a cow as much. The confinement factor of the corral works in the human's behalf, not necessarily the cattle's. However, when working cattle in the open, we have to handle them more conscientiously simply because they can scatter if we don't. This is where "thinking like a cow" benefits both the cattle and, in the long run, the owner's wallet.

Avoiding Stress on Stock

Handling stock takes time; you have to work on their schedule. Nature moves at its own pace. Too often, people are more focused on other priorities and don't always notice when an animal is stressed. Most people see stress in the form of illness or body condition, but they might not recognize emotional stress and how it can affect cattle.

If you've ever been involved with weighing cattle, you're aware of the weight loss or "shrink" cattle experience in just the first 30 minutes after taking them out of their comfortable environment. In a matter of just a few hours, they can lose as much as 5 percent, or more, of their body weight.

While cattle can rehydrate fairly quickly and easily, restoring energy loss in the form of stored energy cells takes a longer time. The savvy cattle owner knows that this time could be better spent storing new cells instead of replacing lost cells.

Same Principles Inside and Outside

All the basic principles that work in the A pen are used outside when you work cattle that aren't confined by fences. You still work off the flight zone and balance point, but outside the pen you have more time to rate the cattle. Because you have more room, you can take more time stopping and turning with a young horse.

If you want to speed up the cow, you just get tighter in on her flight zone and stay closer to her balance point. Most of the time, when you are in the rodear, you want to keep things quiet and reduce stress on the cattle, so you often stay farther outside the flight zone because you want to keep the cattle slow.

Any time we remove calves from the cows we cause stress on both of them. Obviously, when we want to wean calves, we have to separate them, and there might also be other reasons we need to separate them at different times. But when cattle are uncomfortable, they can lose weight, and if they are bawling for each other, they are not comfortable. The sooner they can be back together the better—for both the cattle and the cattle owner's wallet.

"We really didn't have much use for corrals because we were simply used to working livestock in the open."

Let's put a dollar figure on the stress of working pairs. If we have 300 head of 300-pound calves that are worth $1.50 per pound, and we shrink them just 2 percent, that's a loss of $2,700. It's not hard to shrink a fat calf 2 percent on a warm day, not to mention, how the cow might respond in her breeding cycle with a freshly fertilized egg.

Growing up on a family ranch, it was understood that stress on the cattle meant less reproduction for breeders, less gain on calves and yearlings and overall less weight at the scale. Any time our cattle were "violated" or made uncomfortable for any length of time, it was like getting a pay cut, and we were underpaid as it was.

We live in an age where so much is free, instant, and easy. There isn't much value given to preparation and patience, yet preparation is one of the most overlooked ingredients when it comes to working livestock. Cattle that are handled quietly can, with a little preparation, be worked outside without a lot of help or taking the time to move them to a corral.

People tend to put more pressure on cattle in the confines of a corral, which causes more stress than working cattle in open country.

When the conditions are assessed, the plan is handed out to the other cowboys.

Moving cattle to a corral often means calves are moved from their bed grounds, creating unease with their mothers, and then leaving them to travel back to those bed grounds, or recover at a new home. If the cattle are in an area where they're going to stay for a while, and we can work or brand them—or do whatever we need to do—right there without moving the cattle, it is much less stressful on them.

Anytime the cow can stay quietly next to her calf while we sort or brand, it's money in the bank. Any time a person splits the pair unnecessarily, either from neglect or ignorance, this causes stress, and that takes money out of the bank.

Many times I hear people make reference to a "dumb" cow, typically after the cow does something that doesn't agree with the person's agenda. Cows are bigger than we are, so we can't make them do something they don't want to do without using an aid to equalize their weight and strength. We like to think we are smarter than a cow, but how many people can't outthink a dumb cow? It's not about I.Q., so much as patience. For example, letting the cow process the easiest route after we outsmart her by making all the other routes difficult for her takes time.

Remember this fact: If we apply pressure and the cattle don't respond, we have succeeded only in stressing them.

Etiquette on the Open Range

When working cattle, acknowledging the fact that there is etiquette in place, and being responsive to it, gains you points. If you do make a mistake, people probably are more forgiving. Nobody likes a loud, belligerent person around stock, and the stock also responds negatively to such a person. Nothing can humble a person the way livestock can. Livestock has no regard for who you are or what you've done; it's what you are doing now, with respect to the cattle. Most people who have worked much stock realize and respect this. If a person has the right attitude, there's not much need for rules.

When working cattle around the rodear, there are certain rules you follow. Generally, these are unwritten rules and can vary somewhat from ranch to ranch, but the idea is for everyone to maintain respect for the stock and for the other cowboys. Although the rules can change from one area to another, the basic reason behind the rules is to get the job done in an efficient way with less stress on the cattle.

Every ranch has a cow boss on the crew, or if there are several ranches that come together to work cattle, one person is designated to be the range boss. The choice for range boss is usually an obvious one; he's often the one who controls the range that is being worked, or the one who has the most cattle in the rodear. Or he might simply be the individual recognized as the most knowledgeable. At any rate, this person serves as "judge and jury." If someone has a question, or if anyone violates the rules, the range boss has final say in the matter.

"When working in the rodear, there is no corral or pen to hold the cattle that are cut from the herd, so riders simply hold them together in another smaller group known as the 'cut' or the 'pratha.'"

Most people who have worked in any area are aware of etiquette around the rodear, but young riders or those who haven't participated before might not know what's expected of them. Following are some general rules to be followed when working in the rodear:

- Don't ride someone else's horse or use another person's equipment without asking.
- Minimize or eliminate trotting or loping in the rodear, unless you know this is permitted. Many times the range boss doesn't like to see anything but a walk in a rodear because faster horses stir up the cattle, and if you're working pairs, they get mis-mothered.
- If you're roping out of the rodear, you might not be allowed to swing your rope in the herd because this can stir up the cattle. Instead, just bring it up over your head once and throw it, like a houlihan.
- Some ranches want you to catch a calf by both back feet, although this might depend on the caliber of ropers working with you. Catching just one back foot is often discouraged because this makes more work for the ground crew, and can be harder on the calves if you are dragging them.
- Don't ride in front of other riders when going from Point A to point B around the rodear; instead, ride around behind other

If you can learn to throw outside the flight zone of cattle, you can minimize stress not only on what you catch, but also on other cattle in the rodear.

riders. This is the same as proper etiquette at a party; you don't walk in front of other guests if you can walk behind them.

- When you're assigned a position around or in the rodear, don't give up that position, even if you chase a cow back into the rodear on the other side of the herd. You don't stay where you brought the cow back into the herd. You ride back around behind everybody and return to the position you were originally assigned. That same rule applies when gathering cattle for the rodear. If you chase a cow and end up clear over in another area being gathered, once you're done with that cow, you go back to the position you were assigned.
- Don't make people wait on you. Someone always has to be last, but that person doesn't always need to be you. I consider it very rude to disrespect other people's time by making them wait. Again, it's as much about attitude as anything. If you are sensitive to the fact that other people are waiting and you run into a legitimate delay, that's one thing, but being unorganized or lazy is no excuse to waste someone else's time.

There are very good reasons behind these rules. When I was running a crew, I didn't just randomly assign positions in the gather. I considered the horse each person was riding, as well as the rider's experience and ability.

If a person was less capable—either because of the horse he was riding or his own abilities—I put people on both sides of him who were more capable. If someone wasn't familiar with those cattle and that country, I put people on both sides who could come to his aid, if necessary. We didn't just split up and ride along visiting with our favorite person without regard to how efficient and effective we could be about getting the job done.

Stand Still

While we're on the topic of rules, one thing I personally am very strict about is my horse standing still when I mount. This is not particularly rodear etiquette, but simply a safety issue and worth mentioning. I find a lot of riders are so focused on getting on that they don't pay attention to the horse moving around or walking off as they mount. This is like opening the car door and jumping in while the vehicle is rolling. It's dangerous.

If your horse doesn't stand still for mounting, you need to work with him until he stands quietly. Don't ignore this issue or you reinforce a bad habit that makes things more difficult, as well as unsafe.

After the header makes her catch, the calf can be pulled out and heeled.

A Little Philosophy About Ranching and Working Cattle

The idea of using a horse to work cattle is to use a bigger, stronger creature to make our work easier while handling other bigger, stronger creatures. It's difficult for one person to physically handle cattle, but one experienced horse and a rider with a rope can handle cattle much easier. It takes practice to develop the skills needed, but when skilled people and horses put their talents to work, they can get a lot done.

Working with livestock is a very humbling occupation and a never-ending learning experience. It's not something you get into for the money. It's a lifestyle. If you don't enjoy the lifestyle with livestock, you should look for a more lucrative business.

You can't put a monetary value on seeing a young kid excited about roping his first calves, or a young colt working his ears, figuring out how to do his job better.

Your cattle hold condition better if you can work them without stressing them, but there's also a priceless satisfaction in learning and understanding livestock and knowing how to work them well.

Working Cattle

There's always a purpose behind why we cut certain cattle out of the herd first. For example, if we're working pairs, they always come out first because the more people ride through the herd, the more the cows and calves might get separated. If there are bulls in the group and if they are stirring up the rodear, they might be the first to be cut out, just so we can get down to work. This would take priority over different owners who want to get their cattle out first.

Yearlings are usually cut out after pairs, then dry cows, or whatever else might have to come out of the herd. If it's late in the year, and we're trying to get yearlings together for shipping, they are typically taken to another area after they're cut out.

When working in the rodear, there is no corral or pen to hold the cattle that are sorted from the herd, so riders simply hold them together in another smaller group, known as the "cut" or the "pratha." The Charles Russell painting titled *The Roundup* is a good illustration of this practice.

When cutting out pairs and yearlings, they are usually just held in different groups, depending on why they're being cut. If the herd contains cattle owned by different people and we're branding calves, it's typical

The flight zone is minimized to drive the pair from a rodear to the pratha, so they're quiet and ready to stand when you quit driving them.

to "mother up" the calves of the owners with the fewest cattle, and then rope and brand these first. Unless cattle are going to be separated for some reason, the newly branded calves can just stay in the rodear with their mothers.

In a quiet rodear, pairs stay together and are easy to find and cut from the herd.

"It's all in how the cow perceives comfort and relief. The better we understand this about cattle, the better we can adjust so things work to our advantage."

12

Low-Stress Sorting and Branding Techniques

Any time we mix large animals, hard surfaces, mechanical devices, ambition, testosterone, and inexperience, we have more than enough ingredients for things to go wrong.

Livestock doesn't get in a hurry for things to happen in a normal day. We are working with animals that, for the most part, don't have egos. No matter if we work cattle in the rodear or in a corral, we need to check our ambitions and egos so we can be in balance with the stock. This greatly minimizes the stress and panic we inflict on them, and helps us get the job done more easily.

Once the cattle are gathered, they are held until they settle.

Balance Point and Direction

Remember: It doesn't matter if a cow moves toward you or away from you; she responds to your position in relation to the balance point. If the cow is moving the wrong direction, away from where you want her to go, you need to reposition yourself according to her balance point. This causes her to change directions.

We always need to be aware that any time we engage the animal's self-preservation instinct we could cause stress. In many cases, the more stressed the cattle become, the more work we make for ourselves and our horses because stressed cattle move faster, try to escape and try to get back to the herd. The quieter and slower we can work our cattle, the easier they become to handle. We also get the job done faster because every step is controlled in a favorable way.

Watch That Pressure

One thing I've found with sorting cattle is that people often have a hard time waiting for the cow to get prepared to do what the person wants the cow to do. People tend to be impatient and want to make the cow do something right now.

Cattle don't operate on that schedule. If they're comfortable and it's their decision to do something, they work with you much easier. This is why it's often a shortcut in the long run to just sit there on your horse and wait for the cow to find her way out of the pressure you've put on her.

For example, you have a cow sorted out and are trying to direct her to another group of cattle that have already been cut out. She's stopped and doesn't want to run into you, but wants to get back to the herd. You've already turned her a few times and you've maintained the position, so her thought process is looking for some comfort because she's not finding it by challenging you. If you can just be patient and give her a few moments, it might take her 10 seconds to look away from you and see the other cattle in the pratha, the group of cows that already have been cut from the main herd, and decide that's where she should go.

Many times when you sort out a cow, and she does turn away from you, the pressure you put on her still draws her attention. Because she's looking back at you, she isn't focused on the other cattle in the pratha, the ones you're trying to direct her toward. This is where more pressure makes more work and takes more time for you and your horse. Sit tight and just give the cow a minute.

If you've ever had to put chickens back in the chicken house, you found that when you try to push them, they scatter. But if you keep tucking them into the group, you can gather them without making them feel threatened.

The cow is blocked a few times to keep her away from the herd.

When the cow moves away from the herd, I stop and let her feel the relief.

Then those chickens march right into their house. They can't look straight ahead if they feel they need to look back at you.

Cattle are the same way. Instead of constantly pushing on them, which they perceive as a threat, you need to keep setting up roadblocks, so to speak, to guide them in the intended direction. Just keep blocking the cattle from going where you don't want them to go—but don't force them. Then give them enough time and relieve the pressure so they can go where you want them to go.

Like those loose chickens, if cattle feel pressure from you following them, they turn their heads so they can look back at you. When a cow turns her head, she's going to go where her head points. If the cow looks back at you with her right eye, she's most likely going to turn right.

A cow has to experience enough discomfort to want to move away from you, but have enough comfort as she goes away to stop worrying about you. Then she swings her head forward and moves on. You have to measure the amount of pressure you put on her—left and right—so she goes straight away from you, or in whatever direction you have in mind.

Curb Predator Instincts

Too often when we try to accomplish a task, the predator, or hunter, in us comes out. As soon as cattle or horses get a hint of this, they instinctively go into self-preservation mode. The better we can recognize this in ourselves, the closer and more quietly we can work with cattle. If we can't seem to identify the predator in ourselves, we at least need to acknowledge when the cattle or horses try to tell us we affect them in a predatory way.

"Too often when we try to accomplish a task, the predator, or hunter, in us comes out. As soon as cattle or horses get a hint of this, they instinctively go into self-preservation mode."

For example, calves are by nature very curious. If we hobble our horses close to the herd and there are no other distractions, a calf walks up and investigates. That calf smells and tastes the horses' tail hairs and saddle strings if it can reach them. But the moment that calf senses any threat, it hurries back to the safety of the herd and looks for its mother.

We can take a lesson from this whenever we move, sort or work cattle. The faster we move and the more we move directly toward them, the more threatened they feel. On the other hand, the more passive and quiet we are, the closer we can get to the cattle. All we need to do is discipline ourselves and look at the big picture, instead of focusing on going to "that calf" to rope it.

Sorting

On many ranches, it's customary to let the least experienced riders and horses have a chance to sort the cattle first. They can cut out the easier cattle early on, and then the more difficult cattle, or the ones that got away, can be cut out later by more experienced riders on the more experienced horses.

At some ranches, things aren't done this way. You must earn the right to sort cattle. In this case, there might be a couple riders who do all the sorting because they're more concerned about getting the work done than training inexperienced horses and riders.

Personally, I prefer the idea of letting horses and riders learn, and training them. This way someone doesn't always get stuck with the "dirty work" without a chance to learn. In the environment where I grew up, everyone was equal and respected each other. It didn't matter if somebody was the cow boss or a younger kid. When my Uncle Paul ran the big ranch in Nevada, he never asked someone to do something he wouldn't do himself, no matter how dirty the job. That gave people a lot of respect for him.

Always keep in mind when sorting cattle that the goal is to keep things as quiet as possible for as long as possible.

The first step is getting the cattle gathered. Then, once they are together, you need to let them settle. This might take five minutes or 25 minutes, but it's important if you intend to reduce the stress level.

Some calves might have become separated from their mothers during the gather, so you want to give them enough time to find each other, settle down and relax. Cattle often have their own little cliques, just like kids in a schoolyard. When you allow them to settle, these groups find each other, get on different sides of the rodear and relax together. They form their own little segregated areas, even though they're all in the same herd in the rodear. This actually makes it easier to sort the cattle.

Settling the cattle might not take long. The calves might nurse and lie down, and the mothers will stay close.

As we work the rodear, we minimize moving cattle that don't need to be disturbed, and they remain paired up with their calves.

This is never wasted time because, while the cattle are settling, you instruct the help on what they must do and who works where. Some sort out cattle, while others are responsible for holding the pratha(s) of cut cattle away from the main herd. As the cattle are settling, the people involved also use this time to understand the strategy.

Once the cattle are settled, you can begin riding into the herd and start sorting them out. How many riders you have sorting at the same time depends on the number of cattle and size of the rodear. With a large herd, you might have people on different sides of the rodear cutting cattle out at the same time. It also depends on what your working environment allows—the weather, the terrain, how much help you have, how experienced they are, and how quickly you need to be done.

Speaking of environment, if the area you want to work isn't favorable to the cattle, it isn't favorable for you. Whether holding and working cattle in a corral or outside, the cattle tell you some places work better than others. Sometimes it depends on the time of the day, the sun, the shadows, and the wind. Sometimes it's how the people work the cows. It's all in how the cows perceive comfort and relief. The better we understand this about cattle, the better we can adjust to make things to work to our advantage.

Take Your Time

Sometimes cattle work well for only a few hours before they start getting restless and hungry. At that point, they are hard to work, so you're usually better off to just stop, turn them loose and come back to work them later or even the next day.

If you hold cattle too long and they get restless and start fighting you, it not only makes the day's work harder to accomplish, but also makes the cattle sour. Then, the next time you work them, they might get restless even sooner. This goes back to training your cattle, as we talked about in Chapter 6. There's a lot to knowing your cattle and why they do what they do—whether it's good or bad. You need to understand what works to your advantage—and what doesn't—and know what to do about it. Sometimes it's better to pick your battles and come back another day.

If you understand what makes cattle tick, you can let your cattle work with you, not against you. Many times, if you turn them loose to graze and "mother up" for an hour or two, the calves nurse and lay down. Then you can come back, gather the cattle again, work them for several hours and get the rest

Dealing with Ornery Cattle

Sometimes you start sorting and run into cattle that prove difficult to sort. When this happens, rather than make a big ruckus and split up more pairs, just let the difficult cow go for the moment. Leaving her for later saves both time and trouble. Work to quietly get as many of the pairs out of the herd as you can and remember that tough cow and her calf. This way, even if they get separated, you can eventually find both that cow and her calf and bring them out together, even if it's on the end of a rope.

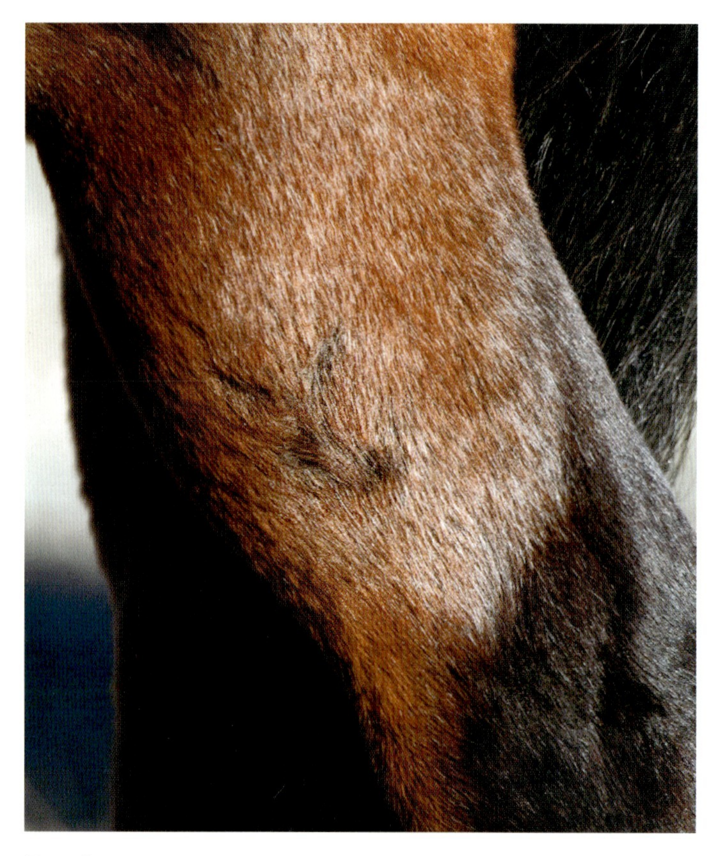

Brands are a necessary identification for livestock on open range, just as a trademark identifies one's product.

of the work done. Giving the cattle a chance to freshen up this way can actually take less time—not to mention less frustration—in the long run than pushing on and never giving them a break. Someone might stay with the cattle to keep them from scattering too far while everyone else has lunch and comes back with fresh horses.

Obviously, freshening up the cattle isn't something you can do in a corral. If cattle are penned, you can't let them spread out, lie down and relax. That's why I like working in the rodear. But even if you work in a corral, you can take some of these other sorting practices and implement them in an enclosed area to make the whole ordeal less stressful on the cattle.

But one thing holds true when working cattle: You can't get in a hurry and have these methods work effectively.

Branding

At one time, branding was the only legal way to identify livestock's rightful owners. On the open range this is still true, and much of the Great Basin area is still open range. Although cattle can be stolen and brands altered even today, it's still easier to identify a cow by her brand and earmarks.

I believe in branding horses for the same reason. Some people don't like the looks of a brand cosmetically. After all, it is a scar. But I hate to rely on a registration certificate, a bill of sale, or another piece of paper to identify a horse. Horses can change colors as they mature and, especially when you have large numbers of horses, it can

As the calf settles on the end of my rope, the heeler moves in on the calf's balance point to set up a heel shot.

Quite often curious calves make an easy target. The key is to stay just outside the flight zone.

be hard to identify them all. The important fact is that a brand always identifies an individual horse to its owner.

As long as an unbranded calf is with its branded mother, the calf can still be identified. If that calf gets too old or gets separated from its mother at some point, then it can't be identified to its mother or its rightful owner. This is why branding generally takes place before the calves get too much size; you don't want to run the risk of a calf losing its mother for whatever reason.

If cattle aren't handled much, there's very little reason for a cow and her calf to get separated. But the way many cattle are handled, the cows get stressed and become separated from their calves. Sometimes the calves don't get back to their mothers, and then you have what are called "leppie" calves, or orphan calves; you can't tell which cow is a calf's mother.

Most people, including me, don't like to brand a calf when it's too young because a calf is more vulnerable to the stress. Some people think a calf is less stressed by branding when it's young because the animal shows less emotion, but I think it is harder for a calf to overcome stress at a young age.

When it's born, a calf doesn't have any fat or stored energy. After a calf is at least a month old, it starts to gain some fat and to store some reserve energy. At this point a calf can better handle the stress of branding.

You want the calf to be strong and healthy, and more than a month old, but when a calf gets to be about three to four months old and 350 pounds or so, branding is more work for the horse and more challenge for the ground crew. You definitely want to brand your calves by this stage, especially if they're on open range.

Time the Horse's Feet with the Cow's

A cow acknowledges the rhythm of the horse's feet, or cadence, in relation to her own. If the horse moves faster, a cow might get nervous and speed up, or want to hide. If a horse travels slower, she might become more relaxed and desensitized. Controlling your horse's feet—not only the speed, but also the length of each step—can make a big difference while working in the rodear.

"Always keep in mind, when sorting cattle, that the goal is to keep things as quiet as possible for as long as possible."

The horse has longer legs than a cow, so if you need to outrun a cow, you often can do it at a walk. If the cow goes the wrong direction in the rodear and is in a comfortable walk, you can get your horse to reach out and take long strides to gain on the cow until you reach the cow's balance

When kids grow up on ranches, they often make the best help. Experience is the best teacher.

point. This can be done without alarming the cow you are working and causing her—and possibly the other cattle in the rodear—to speed up.

On the other hand, you can travel the same speed as the cow, with your horse taking twice as many steps that are half the stride of the cow's step. Because of the difference in activity, the cow speeds up.

Understanding how this works, and having a horse on which you can control the length and speed of each stride, proves very valuable when sorting cattle. If you can out-travel the cow by extending the horse's walk instead of trotting, and then take two or three quick trot or turning steps at the right time to bump up the cow's speed, you have the tools you need to get a lot of cattle sorted without having to work your horse very hard.

Working on Nervous, Young, Inexperienced Horses

If you're riding a green horse, whether in the rodear or at a branding, make it a point to stay farther out of the cow's flight zone because you can't be as accurate on this horse as you can be on a seasoned, more experienced horse. The farther away you are, the more room you have for error in the flight zone and balance point.

Cattle definitely pick up on it when your horse is stressed and nervous. As you try to get close to a cow, whether to cut her out or rope her, she notices when your horse moves quicker than she's moving. If the cadence of your horse's feet is faster than the cow's cadence, she feels more threatened. When your horse's cadence is slower than the cow's, it can calm her. When the horse and person are relaxed, it's easier for the cow to relax. Basically, if you can control your horse well, you have much better control of the cattle.

Branding Rodear Style

Ideally, it's best if you can brand the calves in the same area where they've been living. The less you have to move them, the less stressed they become, so they do better if you can keep them on their home bed grounds. This way, when you're done branding you can just ride away and the cattle can go back to grazing where they've been. Life settles back to normal for them right away.

I think many people overlook the stress placed on cattle when they are driven a long distance and change pastures just for the convenience of getting to a branding corral. Bottom line, stressed cattle mean lost weight and that affects the owner's pocketbook. Having the calves off the cows for four or five hours is enough for them to shrink 5 percent or more. Just standing in the alley, cattle shrink, and it doesn't take long for a calf to shrink 5 percent. With a couple hundred calves, this can easily amount to several thousand dollars.

If you brand rodear style, you don't bring the cattle to a corral, you don't separate cows and calves, and you don't run the calves through a chute or up and down an alley.

It's best to build your fire and catch calves one at a time to brand, doctor or do anything else you need to do. If your help has some training and your cattle have a little training, accomplishing this in the rodear is a very easy process. The calf is stressed only a minute or two, and is then back to its mother.

Often people set up a branding corral and think smaller is better. Smaller means they can go faster without going as far. But as a result, the cattle are violated more and become stressed. This also can be dangerous with too many ropes in a small area.

If you work in the open, you can have 10 ropers with 65-foot ropes, and nobody is in anyone's way. Yet two or three ropers in a small corral can be bouncing calves off the fence or other ropers. Half the time is spent avoiding getting tangled in a rope, whether you are roping or working on the ground.

Social Gathering

There was a time when you found only family and ranch employees at a branding, but today this necessary chore has literally become a social event for many ranches. It's an opportunity for friends and relatives from town to come out and take part in the activities. If they're handy on horses, they can help get the work done, but just about anybody at any age can participate in some way. It's a learning event for the young riders and kids. Even a 5-year-old kid can bring out the vaccine gun and carry the branding iron back to the fire.

Keep in mind that if you invite people who don't understand the importance of the procedures to participate, the safety and quality of work might be compromised. The best way to have a well-balanced experience for both the people and the livestock is to have some discipline handed down by someone who understands what can go wrong, and knows how to keep things going in the right direction.

If you do this, it's absolutely possible to have a safe and educational experience for the more inexperienced help. You also can avoid stressing and violating your cattle, and improve your horsemanship by being considerate of your horses and not asking for more than they can handle.

Brandings can be fun, social gatherings, where wives, kids and neighbors help the regular crew do a necessary job that otherwise would be a chore for those few people.

FEATHERLITE

"You want the cow to associate the trailer with relief from you."

13

Low-Stress Trailer-Loading Techniques

If you work around livestock much today, there might come a time when it's necessary to load a sick or stray cow into a trailer, even if you don't have anyone there to help.

When you're dealing with a single cow, you can always rope her and pull her into the trailer. The picture changes, however, if you have several cows to load in an open pasture several miles from the nearest corral. By the time you get the first one loaded, the other cows could be long gone.

When the cow loads, I don't rush to the door, but just ease in steady and close it. If she gets out, I go through the procedure again until she's comfortable staying in the trailer.

After I pull some air out of the cow, I point her toward the trailer and vill back off ᵗ her feel ᶠ when ads

I have found it to be faster and less stressful for the cattle and my horse to teach a cow to get in the trailer on her own without roping and dragging her, or trying to use a fence to corner her and push her in. I use a low-stress trailer-loading technique that not only is useful when managing cattle, but also can be an exercise in sharpening anyone's stockmanship and horsemanship.

I've often found myself working cattle alone in vast open country, so I've had to learn practical ways to function independently. That's where I learned the principles of working cattle on horseback, such as this trailer-loading technique. It applies the elements of flight zone and balance point, as well as offense and a little bit of defense. The more I can maneuver my horse, the easier it might be to influence the cow, but proper timing of pressure and relief, and the proper position on the cow are what influence her to do whatever I want.

Select a Location

Before there were fences and trailers, ranchers had to know their cattle well and train them to be worked. The cattle were wild; they'd jump up and run when they saw a horse and human coming toward them. If the cowboy didn't get the cattle stopped and settled, they could get away and he'd have to spend several hours, or even days, getting them back together again.

Today, cattle are bred with calmer dispositions and they're handled more frequently, so they're accustomed to seeing people. However, when you work your cattle fast or crowd them, you likely end up with wild cattle that try to get away whenever they see you coming. Instead, it works best to control cattle by acknowledging and using the flight zone and balance point to direct their speed and direction of travel, respectively.

Location can be important when loading cattle in a trailer. It's preferable to work in a flat, open area, free of trees and other major obstacles, and with decent footing. You should be able to ride a large circle around the truck and trailer, so you can gradually work the cattle closer to the trailer.

When you work cattle that haven't been loaded into a trailer or are afraid of the confinement, don't park near a fence line. If you do, as you put pressure on an animal as it tries to leave, you might cause it to run through the fence.

If there are trees or obstacles in the area, you must slow down and decrease your pressure while you get around them. Then the cow learns to seek the relief provided while you negotiate the obstacles, and starts to hide there.

If you put pressure on the cow for trying to leave and you get close to a wire fence, then back off to prevent her from running through the fence, she quickly learns to stay close to the fence. Then, when you apply enough pressure to get her off it, she could go through the fence.

On the other hand, if there's no fence, and the trailer is the only obstacle around, the cow learns to seek relief at the trailer when you apply pressure, encouraging her to move toward the opening.

Help the Cow Find the Trailer

Once your trailer is parked in a desirable place with the door wide open, you're ready to begin the loading process. For the sake of simplicity, we look at how to load one animal here, but bear in mind that you can use this trailer-loading technique on one cow or a small herd of cattle. The cattle might not all load at the same time, but you can simultaneously train the herd by putting pressure on them when they try to move away from the trailer, and then offer them relief when they look or move closer

There's a point when the cow wants to avoid or leave the trailer, and this is when I start to put a little more pressure on her.

to the trailer. Eventually, they seek relief at the trailer, individually or as a herd, and load into it to get away from your pressure.

To work a single animal, begin tracking the cow toward the truck and trailer using offensive tactics, as detailed in Chapter 8. Keep to the edge of the cow's flight zone to maintain forward motion and stay on her balance point to keep her heading in the right direction. Don't hurry! As long as she's moving toward the trailer, you can take it nice and easy.

Once you approach the trailer, continue using offensive moves to direct the cow around the entire rig. Whenever she attempts to leave, ride toward her balance point to circle her around the rig. Don't force her, but wait for her to yield to the pressure and find relief by moving back toward the trailer.

You need to be able to read the cow's behavior and determine when she's getting tired. Then offer her relief by easing your pressure whenever she's near the trailer. The cow most likely sets the pace, but try to keep her at a trot or slow lope, if possible.

A wilder cow moves fast, and you need to take her on a larger circle to get her tired. To regulate the cow's speed, position your horse in her flight zone, staying slightly behind the balance point. Back out of her flight zone when she makes a move toward the trailer. You want her to associate the trailer with relief from you.

It might take three or four circles around the trailer to wear the cow down to a point that she starts looking for some relief. You know she's getting tired when she tries to slow down. Keep in mind that an overly exhausted cow urinates frequently and switches her tail. Avoid driving the cow to this point, or getting her angry or snorty, because she sulls and make poor choices, and you most likely aren't able to get her into the trailer without force.

If you do get the cow too tired, just point her toward the trailer, then get out of her flight zone and allow her to rest and catch her air. Then begin to slightly increase your pressure, gradually riding into the flight zone and on her balance point, directing her closer to the truck and trailer.

-veral approaches before the cow actually gets in the trailer.

When the cow isn't thinking about the trailer, I increase the pressure.

Your first goal is to help the cow find the trailer. After she experiences some increase of pressure for trying to leave, push her toward the trailer. If she tries to leave, she runs into the pressure from your horse and speeds up. Maintain the pressure until she starts to tire, and then turn her toward the trailer if she hasn't already volunteered.

The basic principal is that the cow puts pressure on herself by pushing back on you, but gets relief by moving toward the rig and eventually, through the trailer door. The instant she moves toward the rig, stop your horse or turn him away, and allow the cow to rest. As soon as she moves away from the trailer, ride back toward her and start circling her closer to the trailer again.

It's better to maintain pressure in a position on the cow until she moves away from you. If you reverse and circle the opposite direction, this is fine because you maintain the pressure and position. What can work against you—and favor the cow—is going head-to-head, as a cutting horse would work, and trying to hold her at the trailer door. What generally happens is that the cow feels more pressure in the turn and a horse is late in the turn, which gives the cow some relief. Then the horse catches up to the cow for another turn, and the same thing happens again. In this case, the cow begins dictating the turns, and her relief is associated with "beating" the horse in the turn. Instead the cow should be getting constant pressure from the horse and the relief that comes when she moves toward the trailer. This teaches the cow that when she moves near the trailer she gets relief, and when she moves away from it, she must work harder.

> **"The more I can maneuver my horse, the easier it might be to influence the cow, but proper timing of pressure and relief, and the proper position on the cow influence her to do whatever I want."**

At first, the cow might go around the truck and try to hide under the gooseneck, between the truck and trailer. It's fine to let her experience relief next to the trailer, but don't allow her to get under the gooseneck; she might stay there and hide from your pressure, rather than loading inside the trailer. It's better to ease her away from that area and toward the trailer door.

"The basic principle is that the cow puts pressure on herself by pushing back on you, but gets relief by moving toward your rig and eventually, through the trailer door."

Direct the Cow Toward the Door

Now that the cow understands that the trailer is a pleasant place and leaving is more work, hone that knowledge by showing her that the trailer door area is a better place by moving her toward that spot. When she moves toward the trailer opening, stop your horse or turn away. You might even turn so your horse's tail faces the cow to offer more relief.

Now work up to easing her into the trailer. Once she finds the opening, she wants to inspect it. Give her the opportunity to do this. It might take five or 25 minutes for her to load the first time, but once she figures out that you leave her alone when she's inside the trailer, subsequent attempts become faster. Don't be in a hurry to close the door. Let her realize the trailer is relief, and then ease up to close the door. If she comes back out, just work her some more.

It's not natural for a horse or a cow to want to enter a confined space, such as a trailer. Whenever I deal with fear in a horse, I first try to identify what the horse sees as confining and either eliminate it or show the horse how to accept the confining situation. As with horses, cattle are herd animals, which means they seek out other cattle for security. When training a cow to load into a

As the cow moves closer to the door, I turn away to let her realize more relief as she approaches the opening.

trailer, you have to reprogram her to be like a ground squirrel and run to a hole, rather than the herd, for security.

Having said that, some cattle jump in the trailer right away, while others need encouragement. When the cow does load, make sure you stop your horse or turn him away from her so she feels a noticeable moment of relief. This is often a situation in which turning your horse's tail to the cow is the most efficient and least threatening way to show her relief in the trailer.

The cow might spook at the trailer rattling and come back out. If she's panicked, allow her to exit on her terms, rather than trying to hold her in there. The more pressure you apply in this situation, the more fearful she becomes, and loading in the trailer becomes a negative experience for her. Instead, wait for her to come out of the trailer and then work through the loading routine again. It shouldn't take as long this time for her to find the trailer door. Ideally, you want her to stand quietly inside before you close the door.

However, if the cow tries to exit the trailer and isn't panicked, you can assume a defensive role and ease toward her, gently blocking her path out of the trailer.

With a little experience, the cow seeks the relief inside the trailer, and you no longer have to worry about loading cattle in an open, remote area without corrals.

I keep offering the cow relief at the door so her thoughts of the trailer are good ones.

"If you can control and position your horse properly, he can be a great help to control and position the rope, and, of course, the rope helps control and position the cow."

14

Ranch Doctoring

"Ranch doctoring" is a term many people are familiar with, but the practice of "doctoring" an animal after it's caught is something I am not very familiar with. I couldn't say how many cattle I've laid down solo, or with a helper, for different reasons, but the number that actually were doctored would be a fraction of 1 percent. Although my roping might not be to doctor an animal, we refer here to this as ranch doctoring.

Evaluating the health and strength of the cattle before they are caught might influence the method chosen to catch one.

Some circumstances allow me to get an animal caught quietly.

Personally, my reasons for roping and laying down cattle are more for branding, investigating ownership, and moving an unwilling animal from one place to another, such as a different herd, corral, trailer, etc. I also rope a cow and lay it down for training purposes for my horse, or for the cow, so both can be better prepared the next time a situation might arise. Basically, I am referring to a practice wherein the cow is in a controlled position, whether she is tied down and left, or the rope removed and the cow released.

You've already learned that you and your horse can make a formidable team when it comes to controlling cattle without the convenience of corrals and fencing. Now you can take those skills and meet the challenge of laying down cattle, even if there's no one else available to lend a hand.

Laying a cow down doesn't require a header and heeler to rope the cow; with the help of a confident cow horse, you can handle this task on your own. One-man doctoring is an essential skill for working cattle on the open range and in a cow-camp

In other circumstances it might be easier to set up and catch a calf quickly with a chase.

situation, where you might be working alone. Doctoring also is a great way to sharpen your horse's rating, roping, pulling and overall handling skills.

In very recent years, ranch doctoring has evolved from a chore to an artistic, competitive event in ranch roping, ranch-horse and ranch-rodeo competitions. Actually, the popular sport of team roping has its roots in ranch doctoring. In team roping and ranch rodeo, the key to winning is posting the fastest time, whereas in ranch-roping and ranch-horse contests, you're judged on your smoothness, efficiency and finesse at roping, laying down or controlling a cow on your rope.

When working cattle, focus on efficiency rather than speed. It's when you add speed that you often lose smoothness and finesse, as well as sacrifice good horsemanship. Furthermore, you risk injuring your horse, the livestock and even yourself.

Some people think that without a chute or corral in which to work, nothing can be accomplished with cattle, but that's not true. I've found that working cattle in the open is far less stressful on the animals, and it allows me to hone my horse's skills. It takes experience to predict what can happen, but I can complete the job myself and without physical forms of confinement. I just have to be patient, read and anticipate the cattle's next move, and wait for an opportunity to present itself.

Different Circumstances, Different Challenges

Part of the fun of using cattle to train horses is that each experience can present different circumstances that challenge you and your horse. When you approach the herd, the cattle might scatter and run, or they might remain quiet and compliant. A cow on the end of a rope might submit to the pressure and lie down easily or flail around like a marlin. Whatever situation arises, you need to analyze it and apply the best possible technique to get the job done, even when you work solo.

For example, if you suspect that the cattle are going to leave an area where you plan to work, slow down and take time to settle them quietly. In the long run, you should be able to get more done in less time and with less stress on your horse and the cattle.

"It takes experience to predict what can happen, ... but I just have to be patient, read and anticipate the cattle's next move, and wait for an opportunity to present itself."

If two ropers are working together, they can head and heel the animal.

Desensitize a Cow to Doctoring

There are many ways to approach, rope and lay down a cow to be doctored, and it takes practice to efficiently do the job. There are different ways to get an animal down, but to keep her down you must keep her hind feet off the ground. A cow gets up with her hind feet first, as opposed to a horse, which gets up front legs first. Therefore, if a cow thinks she can get her hind feet beneath her body, she might fight to get up.

To immobilize a cow, you must keep her from thinking she can use her hind feet. This requires laying her down and rolling her toward her back, where she can't get her feet beneath her to get up. Your horsemanship plays a big role both in accomplishing this task and developing your stockmanship skills. The handier your horse, the better he puts you in position to rope and handle the cow, and the better able he is to keep the rope tight and pulled at the proper angle to keep the cow down.

With the cow in this position, it's easy to get close and tie off. I can tie some legs to do what's needed and pull out the tuck in my rope to release my dallies. I then remove the neck rope, holding down the animal by bringing the tail between the hind legs and pulling it tight in the flank. If necessary, I can roll the animal or release the tail to let the animal get up.

As a horseman and stockman, you must realize that when working cattle, no two incidents are ever the same. Therefore, the biggest asset you can possess is the ability to anticipate what can or will happen before it occurs, and choose the best response for the circumstance. If you have only one cow in the pasture, that you handle only once before shipping her away, you can just run up and rope her as though you were in a rodeo competition. However, when you doctor bulls or replacement heifers that you'll have around for a while, you want to do things more quietly and gently so the cattle become easier to handle next time.

On my own, I've had to doctor bulls that weighed more than 1,500 pounds. If I did get one roped and restrained, I could have trouble keeping him down, even with a good horse. I soon realized that I could get the animals doctored more efficiently and with less stress if I could leave them standing up. To accomplish this, I took time to desensitize the bulls without getting off my horse; this way I didn't lose control of the livestock or put the animals or myself at risk for injury. The routine proved beneficial on the cattle that would be around long-term, and the process tended to get faster each time I handled the cattle.

This is why I recommend that you ride quietly around the cattle until their self-preservation response relaxes, or their flight zones get smaller. At that point, you can get close enough to efficiently rope an animal without having to chase it. If you have to chase an animal, it's a surefire sign that you're in its flight zone. If you work in slick or brushy conditions, it might be more efficient to hold the cattle in a fence corner, if possible, or as a herd if you're in the open, and rope them one at a time. It should take only two or three riders to hold up and catch some cattle. Keeping the cattle together poses the least stress on the animals and your horse, and also saves time.

Restraining the Cow

One different low-stress stock-handling technique is to throw a long shot with a rope at an animal on the edge of the herd. This way you don't trigger the cow's flight zone and cause the herd to scatter and run. Remember, the more you engage a cow's flight zone, the faster she moves and the more she tries to stay away from you.

My rope can be placed on the cow's hind feet and tied off again. When my horse is facing the stock and my rope is tight, much of the risk is eliminated.

Another method for catching cattle is to turn your horse's tail toward the cow, as discussed in Chapter 8, and rope her by throwing behind your stirrup. The cow is less threatened because your horse is facing away from her. You also can get closer to the cow by side-passing or backing your horse into position. Quite often, as you ride past the cow's balance point and on the edge of her flight zone, she looks at you, making her head an easy target to rope.

If you have to come up behind a cow to rope it, a trick that can work for getting close is to softly enter the cow's flight zone, sending her away from you. There's an area straight behind the cow where she can't see, so she tips her head to the side or turns slightly to see you behind her. If you change sides, the cow changes eyes and looks back at you. There's a lapse between her changing eyes called the "blind spot." The more worried the cow, the more quickly she changes eyes, whereas the softer or more trusting the cow is, the longer she allows you to remain in her blind spot.

If you wait until a cow softens somewhat, you can use this to your advantage to get closer to the cow. If she looks at you out of her left eye, try to encourage her to change eyes softly by moving behind her, into her blind spot. Stay quiet, as she starts to become more trusting toward you. When you think you're as close as you're going to get with her staying calm, quickly move toward her. In the second it takes the cow to switch eyes, you can hurry and ride a few strides closer toward her.

In some situations, you might have to chase her, but you have a head start and should be able to catch her quickly. In other cases, you might be able to close the gap between her and your horse to the point you can rope her before she has a chance to take off. This can also give a young horse a chance to improve his tracking abilities.

Keep Stress Levels Down

It's less stressful on a cow to rope her from a distance outside of her flight zone—whether she's in the open or within a herd—than to chase her. Keeping cattle in a herd is especially effective when working solo on the open range, where you don't have the aid of corrals, chutes and fenced barriers to contain the animals. If there is any excuse to bunch the cattle to hold or trail them, this can create a good opportunity to get close using the security of the other cattle to keep your "target" calm.

If I can get an animal to stand quietly enough for me to touch it while it's on all four feet, I can administer an injection from my horse.

After touching the cow with my foot, I might pull a few hairs to simulate the prick of a needle. Then while holding my rope snug to the cow's neck, I might penetrate the needle, then attach the syringe and give the injection.

Injection while Horseback

I like to desensitize cattle to allow me to ride up behind a cow and give her an injection from horseback. Taking time to do this makes it easier for me while using my horse to hold the cow and keep me safe. It's also less stressful on the cow than wrestling her down and holding her.

If the animal is already sick and weak, this process usually doesn't take long. If the problem is foot rot, or something that leaves a cow strong and fresh, desensitizing the cow can take a lot longer. When you're by yourself and the stock is big enough that you could have trouble holding an animal down alone, it might be safer and easier to use this method.

To implement this desensitization process, rope a cow and dally your rope, working the slack out as you side-pass up to the left side of the animal, until your horse's head is even with her left hip. This puts you close enough to slowly take your right foot out of your stirrup and gently press on her hip with the toe of your boot. Your goal is to accustom her to being touched at the injection site.

When the cow accepts you being close to her and touching her with your foot, slowly reach down and rub her hip with your hand. Next, poke her with a fingernail, or pinch her, and pull a few hairs to further the desensitization process. She might try to leave, but you have her dallied, holding the rope with your left hand or under your thigh.

After she tolerates your touch in this manner, stick the needle in her, but without the syringe. The next time she stands, she's ready to have you attach the needle to the syringe and inject the medicine. Furthermore, your horse is trained to put you in position to perform the task.

When you finish and need to remove your rope, just use your foot to rub the cow while working closer to her head. As you move closer, keep slipping the slack in your rope around the saddle horn to maintain some tension. This way, if she tries to leave, the rope comes tight; when you're close enough, get hold of the honda and let the dallies go.

I've found it very useful to have a string attached to the honda.

When removing a rope from an animal, the string attached to the honda hangs where I can get hold of it when the rope is still tight.

When I take hold of the string with one hand, I release the dallies with my other hand and pull off the rope without laying down the animal or dismounting my horse.

When a weaker animal can be circled, the rope from the neck loop comes along the calf's right side, then behind it and above the hocks.

The tail of the rope then comes along the calf's left side to the front and eventually lays below the knees.

I then bring the rope along the calf's right side. When the rope is dallied and pulled behind the calf, its front legs step over the rope, and the rope goes between the calf's hind legs.

Ranch-Doctoring Routine

Whether I tend cattle on the range or use my ranch-doctoring skills in the arena, I usually follow one of two methods. Here is the first.

Once you get the cow roped, ride around to her right side, running the rope along her body. If the cow is standing or moving slowly, ride a circle around her. As you circle her, hold the rope off the ground and lay it from the honda down the side of the cow, across the hamstrings of her hind legs, and then bring it around the other side to the front and above her knees.

As you lay the rope on her hamstrings the second time, dally and turn while trotting away from her, pulling the rope tight. When the cow steps out of the rope with her front legs, the rope pulls tightly on her hind legs. If you are too slow, she can get slack and step the hind legs out of your trap. But be quick enough and the hind legs hold the rope; as you pull her backward, she goes down. When the cow is no longer struggling, hold the rope tightly with your hand, remove your dallies and get your horse as close as possible, raising the cow's hind feet off the ground.

If the cow is moving or you can't ride a circle around her, try this second method. Flip the slack between the honda and your hand, so that the rope goes between the cow's front and hind legs. As the cow steps with a hind foot and picks up the rope, it creates a half-hitch when she steps down. Whether she puts one or both hind feet in the half-hitch, your horse can pull her down by dragging her backward.

Once she's down and stops struggling, hold the rope tight with your hand, or let the horse maintain tension on the rope as you circle the cow. Bring the rope up over her ribs so your horse is pulling at her withers or just behind them, which allows you to roll her onto her side. This angle allows you to roll her toward her back and hold her by the hind leg.

Next, get as short on your rope as possible. Put your horse's front feet next to the cow's back, and then get off the horse, standing at the cow's backside. Kneel down over her, pulling up one of her top legs, or her tail between her hind legs, and roll her toward her back to restrain her. At this point, you can tie one or both hind feet to the front foot with a piggin' string, or you remove the head rope, place the loop

By pulling slightly to the side, the calf can be rolled so it lies flat with its feet off the ground.

Then holding the rope tightly, I can ride close, dally and tuck the slack under the tight rope and step off my horse.

With the animal standing or moving, after it's been roped by the head, the slack can be flipped under the belly.

As the animal goes forward, stepping over the rope, my movement creates a half-hitch. I pull the slack tight when the animal steps ahead.

With either both feet, as shown, or only one foot in the dally, I can pull the animal backward.

After the animal goes down, I hold passively until the animal lie quietly.

Next, I flip the rope from the hind feet over the cow's withers.

I then roll the cow to get the feet in the air and wait for the animal to quit struggling.

around her hind feet and possibly secure the rope with a half-hitch. The rope is dallied to your saddle horn, so when your horse pulls the slack out of the rope, the cow's hind legs are immobilized and she can't get up.

If both hind feet are in the loop, the rope is as short as possible, pulling the hind feet up so you can tie off. Then you're free to do whatever you need to do while your horse holds down the cow.

When you're on the ground, you should be close to your horse and the cow, so you only need take a step or two if you need to help. Your reins should be tucked out of the way, under the headstall. If you have a mecate, it should be tucked in your belt so it's out of the way, but within easy reach.

Staying Safe

Any time you work with large animals, you must acknowledge that their size alone can get you hurt. Should they desire to use their strength and speed against you, the danger further increases. If that's not enough, add a stout rope to the equation and tie everything together.

If you remember only one word to keep you safe, it should be "control." This comes back to good horsemanship. If you can control and position your horse properly, he can be a great help to control and position the rope, and, of course, the rope helps control and position the cow.

"The handier your horse, the better he puts you in position to rope and handle the cow, and the better able he is to keep the rope tight and pulled at the proper angle to keep the cow down."

Having control of your horse and the rope is essential to prevent it from burning, slapping, grabbing, or cutting off your

By bringing the tail between the hind legs and pulling it tight in the flank, I can hold the animal to keep its hind feet off the ground so it can't get up.

fingers when working with an angry cow that can do much damage. If your horse is facing the stock and your rope is tight, much of the risk is eliminated. Also, always keep your eye on what you've roped. If at any time you're not confident you have control or if you find yourself in an unsafe position, there's no shame in turning your rope loose.

Dallying might be intimidating to some ropers who don't understand the basic principles of controlling the situation, or who don't have an experienced eye to keep things from getting out of control. One benefit to dallying is that you can slip your dallies, giving your horse confidence that he can hold anything. You never have more pounds of tension on the rope at any given time than your horse has the confidence to handle.

Let your dallies slip, and by the time you get to the end of your rope, your horse might have traveled 100 feet and pulled the cow 50 feet. As the cow weakens, the horse gains the strength and confidence to pull more weight. Eventually, he pulls the cow stride for stride.

Functional Feature of the Wade

Tom Dorrance and Cliff Wade designed the Wade saddletree with the intention of enabling a cowboy to control his dallies from the ground by tucking the tail of his rope underneath itself where it passes over the saddle fork. This is handy when you need to get off your horse to doctor a cow; you can release your dallies when you're on the ground by pulling the little tuck from underneath itself on the lip in front of the saddle horn.

When you're ready to loosen the rope, you have the coils with you on the ground

With some dallies around my horn and the rope snug against the lip on the Wade tree, I pull a tuck between the lip and the rope going to the calf.

My coils come down the left side of the horse's neck with me and when I'm ready, ...

and pull the tuck loose from the saddle fork lip and release the dallies. Then you can reset your rope on the cow, or do whatever needs to be done, and also pull the rope tight. There is only one dally, at most two, around the saddle horn that you can pull from the end of the rope in your hand to remove most of the slack. Or, when your horse backs, he pulls the rope tight. Once the cow is immobilized, you can tuck the rope back under the saddle horn or tie off with a more solid method.

Improve your Roping Know-How

Roping doesn't have to be dangerous—it can be challenging, useful, fun and addictive, when practiced properly. If you want to find someone to help you improve your skills and safety, find a person who first enjoys roping and horses, and is fairly competent with both. If that person has some worn-out ropes and horn wraps, five digits on his or her roping hand and a heavy callus on the little finger, he or she probably knows something about roping and can help you achieve your goals.

... I can pull out the tuck and release the dallies.

"Becoming a good roper always takes practice and experience."

15

Ranch Roping for Work and Fun

Just like most rodeo events and other competitions for cowboy participants, ranch roping started out as a practical application. In some parts of the country, ranch roping is still used for branding and other purposes, much as it has been for the last 100 years. That's how I knew it for many years.

Once ranch roping hit the arena and became a sport, ethics had to be enforced by rules and judges. Any time people take something practical and turn it into sport, they start looking

With a basic forehand swing, as the loop goes over my head and back, my palm rolls upward, and the honda is on the trailing side of the loop.

for shortcuts to win the prizes, and a certain amount of the original application is lost.

I appreciate the ranch roping competitions because they increase awareness of and help preserve some of the old ways. But I also like to encourage people to practice these methods and traditions in a non-competitive setting. In the long term, I think developing these skills actually gives a better, more efficient way to handle cattle, and also allows us to enjoy the culture and share it with friends and family.

"A good horse can make learning to rope a lot easier. ... Just having a quiet horse that doesn't move when you don't want him to move is a great help."

Too often, people just want to get the job done in what they perceive to be a quicker or more efficient way. In many cases, the opportunity to develop practical roping skills simply isn't there today.

That definitely wasn't the case when I grew up.

Then, there was very little discouragement handed down to a youngster wanting to rope. Obviously, there were times when we kids needed to be doing something other than roping, but overall, we always were given a lot of opportunities to learn to rope and always were encouraged to improve.

My granddad Albert Black was one of the best ropers I've ever seen to this day—inside or outside the arena. He taught me sound roping principles and gave me as much guidance as anyone. I think he told me as much what not to do as what to do. He realized that I needed to learn from feeling my rope and the stock on the end of it. That's something that is hard to teach and can only be learned by experience.

Back when I was in cow camp with my older brother, Terry, our cows ran on open range, and it wasn't uncommon to find some of the neighbors' cattle mixed with our own. We boys always were eager to practice roping, but Dad discouraged us from doing so on our cattle. My brother and I soon made it routine policy to hone our roping skills on the neighbors' cows before we returned them to their territory. We figured by sending the cattle home with those fresh memories, they wouldn't feel an open invitation to return quite so quickly.

The Original Californios Ranch Roping & Stock Horse Contest has become popular in recent years. This shot was taken at the 2010 event in Reno, Nevada.

JENNIFER DENISON

There is a wide range of reasons why we need to rope an animal, and a good all-around roper has a wide range of methods to get one roped and handled afterward.

One of our favorite ways to improve our roping skills, as well as entertain ourselves, was to create a little brotherly competition while moving cattle. The rules were simple: We each started at the same time and roped a neighbor's cow. The "winner" was whoever could get the animal controlled and be the first to remove his rope; whatever method it took to remove the rope was considered fair. We often roped mature cows on green colts, so things got a bit "Western" on more than one occasion.

This certainly relieved the routine of cow camp, but it also helped us become better ropers. We also learned more about stockmanship and horsemanship as we used those skills to handle the animals in ways that allowed us to get our ropes off faster than we had before.

Experience Essential

As with everything else about handling livestock, when it comes to roping, there is no substitute for experience. For whatever reason, roping seems to come easier to some people than others. I think a lot of it is just feel and judgment, but becoming a good roper always takes practice and experience, no matter how much natural ability someone might have.

You need the opportunity to get the right experiences if you're going to improve your skills. For example, a good horse can make learning to rope a lot easier. He doesn't have to be a highly trained or competition-level horse. Just having a quiet horse that doesn't move when you don't want him to move is a great help.

If you have to think about your horse all the time, and try to put him in the right position, that makes learning to rope more difficult. But if the horse is on the same team with you and he's familiar with the position where he needs to be, you can concentrate more on your roping and not so much on your horse. A lot of professional rodeo ropers think more about their roping than they do their horses. The cowboys have to do this to be competitive, but to a degree, they sacrifice their horses; cowboys can only do this to a certain point before they begin to sacrifice their roping.

The same principle applies on the ranch. When I ride a young horse and have to

Experience—The Best Teacher

So much of roping is feel and judgment based on previous experience. If it feels like things are going in the right direction, you can just go with the flow. If your horse keeps moving in the right position and your swing is consistent, it all comes together, and you probably make the catch.

However, if your feel and judgment, based on your experience, tell you conditions aren't quite right for the catch, you need to make adjustments. You might need to reposition your horse or wait for the cow to be in a better position, or reposition your rope in order to make a successful throw.

concentrate on getting him in position to rope, I don't catch as well because I'm not concentrating on my roping as much as I am on my horse. I rope better when I'm on a horse I don't have to think about so much.

Roping is just like walking and chewing gum. After a while, you don't have to choose between thinking about your roping or your horse; you can give them both equal attention.

Getting Started

When you start roping, you should practice roping a dummy. Most people are going to have the best luck learning with a soft nylon rope. A soft rope usually is more forgiving and easier to handle than a stiff rope. You don't necessarily have to take roping lessons, but you need to observe, be willing to learn from your mistakes, and put in plenty of practice.

In the beginning, the most important thing isn't catching livestock, but rather learning to handle your rope and getting a good feel for it. You really need to live with that rope. You pull your boots on in the morning and pick up your rope. On your way out to the barn or corral to catch your horse, you rope a few rocks going across the driveway. You continually rebuild your loop and throw it. You keep trying different angles, different swings, and different throws. Pretty soon, you are so familiar with that rope, it becomes an extension of your arm.

As you learn, don't make the mistake of practicing only one way to throw. Many people just swing a forehand and throw the rope only at a target out in front of them, but it's best when beginning ropers don't restrict themselves to just one type of throw. You don't limit yourself to just picking up a coffee cup when it's only in a certain position on the table. The same applies to roping. You want to be more versatile, and you can only become that way by living with a rope and doing a lot of different things with it.

If you stick with roping, you eventually get to the point that you can do things with that rope that no one ever taught you before, or you've never even seen before. Whether it's rebuilding your loop or throwing a loop, you develop your own style of roping, and then you can capitalize on opportunities as they are presented.

Keep an Eye on Your Stock

Always keep an eye on any animal you rope. I recall one time when I was dragging a calf to the fire. Just as I looked ahead toward the fire and ground crew, a heeler moved in, and the calf shot up the rope and to my left. In a split second, the rope went under my horse's tail as that calf hit the end of the rope. If I'd seen that the calf was going to turn and run behind my horse that way, I could have turned my horse or popped off my dallies. But by the time I knew we had a problem, we were already in too deep.

When your rope is tight, you can feel where your stock is. If you can't see the stock, or feel it, you need to pay attention. Don't always depend on other people to look after your safety. Always keep in mind, whenever a rope is involved, things can go wrong fast.

Simple Roping Truths

When you swing a rope, centrifugal force pulls the rope away from your hand. Whenever your hand releases the rope, the thrust makes the rope leave your hand in a straight line. The moment you let go of the rope, you eliminate the centrifugal force. From that point, momentum and gravity dictate where the rope goes.

How you handle the rope as you swing affects the loop you're creating. Your hand can't rotate all the way around like a merry-go-round, so there's a point in the swing where you need to flip your hand. For part of the swing your palm is up, and your palm is down for part of the swing. As your palm turns up, it rotates the loop, with the underneath part of the loop turning over to the topside, and as your palm comes down again, the loop rotates back down. When your palm is down, your honda is on the leading side of your loop. When your palm is up, your honda is on the trailing side of the loop. Basically, the honda travels in the same direction as your thumb.

The point where a loop turns over affects how the loop leaves your hand. If the loop is vertical when it leaves your hand, it continues to travel vertically. If the loop is horizontal when it leaves your hand, it continues to travel horizontally. The exception to these comments is what's called a turnover loop, where you start the rotation of the loop before you release it, and the rotation continues afterward.

Because there are times and reasons you want the loop to leave your hand vertically, and others when you want it to leave horizontally, you have to know which direction you want before you let the rope go.

In addition, the angle, or plane, of the loop is a big factor because the angle affects where the loop is sent. I get into specific types of throws and swings, and when to use them, in the next chapter.

The art of rebuilding a loop should be given as much attention as throwing the loop. Although a lot of people don't pay much attention when they rebuild their loops, the opportunity is there to be more efficient and smoother between each loop that you throw.

There are different ways to rebuild your loop, and as long as it's quick, you have more time to deliver more loops.

One thing you want to make a habit from the beginning: Keep your coils from getting tangled. When you dally, a tangled coil going through your hand easily can catch your thumb and pull it into the dally. To avoid this, get in the routine of recoiling at least the outside coils you plan to throw and dally.

f I want to add coils into my loop, I start vith the honda halfway down my loop.

I roll the tip of the loop out and back, releasing it from my fingers...

... and hooking my thumb in the loop while the honda is back.

The momentum of the loop continues to roll the honda to the front again. My thumb is still hanging in the loop, but now my fingers are on top.

Next, I start to grip the loop with my fingers, bringing them inside the loop.

As the momentum rocks the loop forward, my fingers reach toward my coils to pick up my spoke. With practice, the honda goes halfway down the loop each time and is ready to swing.

The reata is a nice rope to throw long distances.

Which Rope to Use?

The type of rope I prefer to use depends on what I'm doing. When I rope in competition and need to be fast, I like a stiff rope. When I rope a colt in the corral or little calves at a branding, I might want to use a softer rope.

Some ropes are heavier than others, and if it's windy or I'm throwing a distance, then a heavier rope—like a reata—would be my choice. A reata is about as heavy as any rope I can use. It also has less drag, whereas a nylon rope, especially when it's old and fuzzy, catches a lot of air and has more drag.

With my coils on the ground to my left, my rope is always in front of me if I need it. I hold the calf's front leg, pull the tucked rope to release my dallies and remove the head rope.

Controlling the animal by the front leg, I place the loop around both hind legs before I tighten the rope running to my saddle horn.

A reata slides down and underneath the cow's hind feet easier for a heel shot and goes through the air easily. But a reata is more expensive and easier to break than a nylon rope, which is why some people don't like to use them. Once you learn to rope with a reata, it is a real luxury. A reata can't take a hard jerk; you can break one pretty easily that way. However, once there is tension on a reata, it's relatively strong.

A certain amount of maintenance is involved in roping with a reata. When it's new, it's kinky and hard to manage, so you must soften it by pulling it through some holes in some wood and keeping it greased with an animal product, such as liver, tallow or calf testicles, rather than petroleum-based products. If a reata sits, it gets kinky, so you might have to grease and stretch it, and work with it more than you do a regular rope. As you learn more about a reata, you discover how to make it soft, medium or hard. You actually can make a reata as soft as buckskin or leave it stiff as a cable.

Proper Ways to Tie Off

There are different ways of tying off a rope, and I might use a particular way depending on the circumstances, the size of the stock, how experienced my horse is, how long I might need to hold down the animal, and how much the animal might struggle. These are all factors in determining which method I use to tie off my rope.

"Pretty soon, you are so familiar with that rope, it becomes an extension of your arm."

As long as any person is safe and maintaining control of the animal in an efficient way, there's no "right" or "wrong" way to tie off. There are different methods and variations, but the following are the two methods that I use the most. I generally rope

After several dallies, the rope is doubled and the resulting loop is brought under the horse's neck, around the tight rope to the calf, and then hooked over the saddle horn.

The tail of the rope, half-hitched over the horn, holds the rope against the horse's neck and prevents him from going right or left. With the calf's feet raised, the calf can't get up and the horse doesn't come forward. Then I can work while the horse restrains the calf.

right-handed, so these instructions are right-handed. If you rope left-handed, just switch right for left, and things work the same way.

Probably the safest way to tie off is to have the animal you've roped held really short by a hind foot or feet. The rope should be short enough on the right side of your horse that the animal's hind feet are within a couple feet of your horse's chest when the rope is dallied around the horn. Have enough dallies to hold the cow easily without a lot of tension on the rope going to the coils.

After you dismount, take the top coil from your left hand and double the rope. Then take the doubled rope under the horse's neck from the left side and go under the rope, on the horse's right side, that goes to the cow's feet. Bring the doubled coil completely around the other rope and back under your horse's neck.

Now you can draw the loop, where you folded the rope, back up to your horn. Pull the slack from the folded loop back toward your coils and then use the slack to put a couple of half-hitches over the horn to tie the rope solid.

When this is done properly, the horse is not likely to come forward because the cow is there, and the horse can't turn left or right because the rope comes along both sides of

Roping with Bill Dorrance

One time I went to see Bill Dorrance, and after we'd visited a while, he asked if I wanted to do some roping. I thought he needed some cattle roped for whatever purpose, so I said, "Sure!"

We saddled and went to the corral. Bill then drove this little steer out of the corral and up the driveway. He stopped the steer, and we sat on our horses and talked a little while. Meanwhile, this steer just stood there, chewing his cud. Bill shook out his rope and started showing me different shots. To my surprise, the steer didn't even try to leave; he just stood there, blinked his eyes and took it. Bill had raised this steer, and he was pretty much trained to be used for practice.

This was probably one of the few roping "lessons" I've ever had in my life. At the time I knew only the basic *culo* shots. The word, culo, is taken from the Spanish word for "rump," which is what a variety of culo shots target. I was pretty unfamiliar with the numerous culo shots Bill was showing me, and I came away with a handful of new shots to practice and improve. Many times I've thought back on that day and wished I'd had a recording to play back later so I could make more sense of what he tried to teach me.

Another time, Bill was on his way to visit his family in Oregon, and he stopped at my dad's ranch in Idaho and stayed with us a couple days. My brother and I wasted no time in finding a few calves and horses that needed branding. The next few days turned out to be a true educational experience as we got to watch Bill rope in a working application and get pointers from him on our own roping.

Bill was definitely one of the best ranch ropers I've seen. He wasn't a fast roper; his style was to take plenty of time to set up a shot that was going to be effective.

This shot is of Bill Dorrance and me during a branding at the family ranch in Bruneau, Idaho. That's my brother, Terry, in the background.

Anytime there is a question of a dangerous situation…

…open your hand, pointing your fingers toward the horn.

Pull your hand out and get away from the dally. A lost rope is easier to regain than a lost finger, or even a little hide.

One of the most important things to learn while roping is to become comfortable dallying. It doesn't matter if you have a slick horn, use rubber or tie hard. That only means there are different types of wrecks ahead for you.

his neck to prevent that. If the horse can drag the animal backward, he can back for only a limited distance, depending on the size of the cow. Sooner or later, the horse gets tired of backing.

"There are different ways to rebuild your loop, and as long as it's quick, you have more time to deliver more loops."

Another method of tying off on a Wade saddletree is to take some dallies and then tuck the end of the rope between the fork of the saddle and the rope as it comes from the horn over the saddle fork toward the animal. This method allows you to take the coils of your rope with you to the animal, and the horse can hold your rope tight until you have control of the animal. Then you can pull on the loose end of the rope, releasing the tuck and allowing the dallies to slip, which lets you replace or remove the rope from the animal as necessary.

If you reset the rope on the animal, the dallies are still on the horn, and you can either pull the rope tight around the horn or give your horse the signal to back up and he can tighten the rope. Then your horse can take control of the animal again. If you need to secure the dallies in some way, or get back on your horse before releasing the animal, you have those options.

Roping Safety Tips

- If you suspect trouble, keep your horse pointed toward your rope and the stock.
- Watch your stock! Once you rope an animal, keep an eye on it so you know where your rope is at all times. It takes only a split second for a roped calf to change directions, and your rope—with an energetic calf on the end of it—can end up under your horse's tail.
- Keep your rope off the ground once you've caught an animal. You always want to have some tension in your rope; slack can be hazardous.
- The saddle horn is like your nose; it's always in the same place. You don't need a mirror to pick your nose, and you shouldn't look at your saddle horn to dally.
- Keep your thumb pointed up the rope toward your coils and away from the saddle horn while dallying.
- Don't try to grip the rope until you have all the dallies needed to stop or slow down the slipping rope. The rope can't burn you if you don't take hold of it; friction causes heat, and the more you grip the rope, the hotter that rope becomes.

- Keep your coils from getting tangled in your hand. Poly ropes and long ropes especially tend to tangle easily. You don't want to dally and have that rope going through your hand fast—only to have a tangled coil catch your thumb and pull it into the dally. It's a good idea to recoil at least the outside coils when you've been throwing and dallying.
- Don't be afraid to let go! When you try to hold onto the rope, you can get hurt. If you have even a hint of a foul in the coils, simply open your hand with all your fingers pointed toward the saddle horn and then jerk your arm away from the horn. You might have to chase a cow to get your rope back, but at least you still have all 10 fingers to do so!
- When you heel cattle, keep track of where the header is in relation to the calf because you can change the animal's balance point and cause it to change directions.

Help Your Horse

When you rope on a green horse, you can do things to minimize confusion and help him have a good experience. Try to make the horse's first roping experiences somewhat consistent, as this makes it easier for the horse to figure out things and know what he's supposed to do. For example, when I'm on a green horse, I like to hold calves at the fire while the ground crew does their work. This gives my horse a few minutes to relax and digest what's happening.

As he begins to get some experience, the horse can start to anticipate his job when there is a set routine, and his anticipation can get in your way. To discourage this, you can try to take different shots from different angles, position your horse differently, back up, side-pass, and rope from both sides. Also make it a point to dally off both sides to position the calf for a heeler. This helps to rest the muscles in the horse's back, as well as make the work more interesting and comfortable.

Whether you're on an older horse pulling a heavy animal, or riding a colt that is learning to pull for the first time, if you line the rope from the horn over the cantle and straight over the horse's tail, he is able to pull evenly and balance with both sides of his body. This technique helps the green horse learn to pull well and also makes the task easier on an experienced horse pulling heavy stock. In addition, when the horse has a hard pull, slipping the dallies helps the horse keep momentum and not get discouraged. You never want to stop a horse with a heavy pull, so slip your rope before he quits pulling or turns around.

The horse's confidence level is so important. You always want your horse to think his job is easier if he can just keep going. It's a lot better to let a horse think he can pull anything on a rope when you ask him to lean into the load. If you put him in a situation when he discovers he can't pull something, you have a hard time convincing him he can do it the next time you ask.

Holding a calf for the ground crew is a good time for a young horse to relax and evaluate everything going on around him.

"Reading cattle can teach us a lot about reading horses, and handling cattle with horses gives our horses purpose. The purpose of working cattle horseback is to improve the horse and make the job easier; it takes practice to develop these skills."

16

Get Practical: Different Loops and When to Use Them

Dragging a calf at a walk allows the calf to have three feet on the ground pulling back. A calf pulls less in a trot and even less when loping or bouncing along.

I learned early on that the more I roped, the more I could develop a variety of loops that come in handy in different situations. Those loops just add more tools to the tool chest. Starting out, a roper might practice numerous ways of swinging and throwing, but nothing beats practical experience to teach what works best and when to use a particular loop.

Make the Swing

You can swing a rope in any one of several ways, and the particular swing you use helps you set up a particular shot when you throw your rope. Again, practicing the various swings is good, but the best teacher is practical experience so that it becomes second nature to use a certain swing to set up the shot you want to take.

When you rope right-handed and use a **forehand swing**, the rope comes from your right side to the left as it passes in front of you. Your palm is down as the tip of your loop approaches the target. Then, whenever your hand is at or just past the target, your thumb points down. As you bring the loop around and over your head, your palm rotates upward. As soon as you start to bring your loop forward again, turn your palm down by raising your elbow as soon as you can. You want to give the loop as much time as possible for it to become horizontal again when it approaches the target.

If you rope right-handed, a **reverse swing** is the opposite of a forehand swing. Your loop rotation starts behind you, traveling from right to left, and then forward, and the rope moves from left to right as it passes in front of you.

The **houlihan swing** also is a reverse swing with the loop first rotating to the rear, then from right to left, but your thumb is down as the rope travels from left to right in front of you. This means the honda, like your thumb, is down as the loop goes from left to right, but as your loop goes from front to back, your hand rotates so that your palm is up.

A **backhand swing** is a reverse swing that also travels from right to left behind you and then comes forward to pass in front of you. As you swing the rope back behind you, your palm comes up, but when your hand is behind your head, you turn your hand so the thumb points up as your hand comes forward from back to front.

The **turnover swing** can be delivered either **clockwise** or **counterclockwise**. Whatever the case, when using a turnover swing, the tip of the loop needs to point in

*The basic **forehand swing** has the loop horizontal for a delivery from behind. The palm is down as the loop is coming forward.*

The palm rotates up as the loop is going back.

the direction where you plan to release the loop. That way, when you let go, the loop travels to the side of or above the target and then turns over, dropping out of the air onto the target.

A **clockwise turnover swing** is also referred to as a *Del Viento* shot, which is thrown above the target. When the loop leaves your hand, it rotates clockwise one time and then falls straight down on the target while the loop is still rotating. I often find this a convenient swing when I try to rope a calf while it's using another cow as a "shield."

Use Position to your Advantage

If you're going to throw a shot and want the cow to go right to take out the slack, position yourself to the animal's left. When the rope hits her, she turns to the right as she's caught, and this helps you take the slack out of the rope.

*The **backhand loop** turns over as it comes forward...*

...so the honda is on top, on the leading side of the loop.

The backhand delivery is made with the thumb up. A backhand can be delivered vertically or horizontally.

The backhand catch is made with the honda on top of the loop for either a heel or a head loop.

*The **houlihan** has the honda on the bottom when it's at the target, …*

The **counterclockwise turnover** rotates in the opposite direction of the clockwise turnover. The counterclockwise turnover loop goes above or to the side of the target and rotates counterclockwise one time before falling on the target.

One advantage of the turnover swing is that the loop comes down or slightly back toward the target, so this loop can be thrown against an obstacle such as a tree, fence, or another cow. The slack doesn't strike an object because the tip of the loop ropes the target, and the change in momentum drives the slack back up the rope. Used with a heel loop, the turnover swing makes it easy to throw the loop on the opposite side of a calf. Then the loop comes back toward the roper, bringing the tip of the loop under the calf from the side opposite the rider.

… on the leading side of the loop as it comes forward, …

… and as the loop passes the target, the wrist rolls the loop over.

The thumb is down, as is the honda, in the delivery of a houlihan …

… and the catch is made with the honda down.

Different Loops and When to Use Them

Most swings and throws can be used as head or heel shots, but experience teaches you which ones work best in different situations. Let's look at a number of loops and when you might want to use each. For the record, loop or shot refers to the entire process, from setting the animal up to gathering the slack. Delivery or throw refers to the process of sending the rope to the target.

Flat loop: A common head or heel loop, the flat loop starts with the basic forehand swing mentioned above, and the loop is released at the point where it travels directly and horizontally to the target. A flat loop typically is used when you follow an animal straight ahead of you. As the loop falls over the animal's head, the slack in the loop goes to the left side to form a figure eight.

Sidearm: A sidearm loop uses the same swing and throw as a flat loop, except that the sidearm loop is swung and thrown vertically instead of horizontally. A sidearm can be used as a head or a heel loop when the roper is somewhat perpendicular to an animal that travels from the rider's left to his right.

Underhand: Also used as a head or a heel loop, the underhand uses the same vertical swing and throw as the sidearm with the loop delivered much like an underhanded pitch. The underhand loop is handy when the roper is perpendicular to an animal going from the roper's right to his left. This shot, with a slight change in the delivery to turn the loop a quarter-turn, also can be used when the target animal is facing the roper. Although referred to as an underhand in the Great Basin area, this throw also is known as "scoop loop" in the Northern Range regions.

Forehand: A forehand loop is swung at an angle, sometimes as much as 45 degrees, but the angle of the loop always is somewhat consistent with the angle of the target animal's shoulders. The roper's elbow is up and the palm is down. As the animal goes from right to left with the rider perpendicular to the animal, the low point of the swing, where the tip of the loop is directed downward, is on the rider's left side. The loop points in the same direction the cow is traveling, and, again, the loop's angle is similar to the angle of the cow's shoulders. When your horse is pointed the same direction as the cow, the low point of the loop, consistent with the angle of the cow's shoulders and her direction of travel, is in front of your horse.

*The **sidearm** is swung somewhat vertically from the right side of the animal ...*

... and delivered with the honda up.

Rope with a Large Loop and Throw Long

I often enjoy using a large loop because of the opportunities it gives me when roping. When I throw a long shot, it's often beneficial to hold three or four coils in my loop hand. This gives me an advantage because, if I lay the coils in my right hand and hold them flat, I can throw that rope a lot farther because there's no resistance from the coils coming out of my left hand.

The important thing: Be sure to stack your coils when throwing a long distance shot so that the coils leave your hand smoothly and safely.

*The **underhand** is swung vertically from the right side …*

… by leading with the wrist as seen in this photo.

The underhand loop's angle can rotate through the delivery …

… and land angled the opposite of the swing.

Offside: This loop is known by this term because the roper throws off the left side of the horse toward stock on the left, but dallies on the right after he turns his horse to face the cow. The offside is consistent with the forehand throw, but is used when your horse points in the opposite direction of the cow, and the low point of your loop is toward your horse's tail. An offside loop should still have an angle consistent with the cow's shoulders.

The offside shot also can be thrown with a flat loop when the cow points away from or faces toward your left stirrup, looking at you. An offside shot basically can be thrown at the cow from any direction. If the cow points in the same direction as your horse, an offside houlihan shot would be more appropriate because the reverse swing allows the loop to land more accurately.

Houlihan: A houlihan loop functions the same as a forehand or backhand loop but basically is a reverse-swing loop delivered so that the honda is leading as the loop travels toward its target.

The houlihan can be thrown as an offside loop when the animal faces away from or toward your left stirrup. The houlihan and the forehand loops can be thrown in 360 degrees—from your horse's front, back, left or right sides. Depending on which direction the target animal faces, either shot, the houlihan or the forehand, can be either more or less favorable to make the catch.

For example, if the calf's head is at 10 or 11 o'clock and he's parallel to your horse on the left side and facing the same direction, a houlihan is a more favorable throw than the forehand. If the calf is parallel to and pointing in the same direction as your horse, but is off the right side with his head at 1 or 2 o'clock, then the forehand or the backhand is a more favorable throw than a houlihan. If the calf is parallel to your horse and on the right side, but the calf's head is pointed in the opposite direction at 4 or 5 o'clock, a houlihan throw is more favorable than a forehand throw. If the calf is on your left side, parallel to your horse but pointed in the opposite direction, with the calf's head at 7 or 8 o'clock, the forehand throw is more favorable than a houlihan.

Backhand: A backhand shot can be used as a head or heel loop. The backhand loop should be delivered in a vertical position, and this can be accomplished by throwing at any point in the swing when your thumb is up. As a head loop, the backhand throw can be favorable when the calf faces the loop as the rope approaches. The backhand heel loop has a nice delivery as an offside shot thrown from the left side of your horse when tracking an animal on your left.

Heel: The most common heel loop is thrown from straight behind or slightly from the right side of the calf. Regardless of a rider's angle in relation to the calf, the angle of the swing needs to be consistent with the angle of the calf, meaning the the tip of the loop, or low point, needs to face toward the calf's front end or slightly to the calf's left side. If the calf's feet are in the air as the loop is delivered, the loop can be more horizontal than vertical. This delivery is referred to as "roping them out of the air" or a "floater." Whereas, if the calf's feet are on the ground, standing or walking, the loop needs to be more vertical and thrown with a little more force to "lay a trap" for the calf to step into with his hind feet.

Culo: Taken from the Spanish word for "rump," a culo shot is any loop that goes over the target animal's rump, and there are perhaps 15 different culo shots. Depending on the area of the country, this throw might also be referred to as a "flank shot" or called a "hip shot."

A culo can be delivered to either side of the target animal and from the horse's right or left side. If the calf is standing still when the loop hits it, the loop folds in half, with half the loop going over the rump and the other half under the animal's belly, which places the tip of the loop where the calf can step into it. Because the loop falls over the animal's rump, if the calf backs up, the animal still steps into the loop and is caught. When the animal is moving, often the bottom of the loop goes around the feet, and the slack is pulled before the top of the loop lays over the top of the calf's rump. This makes the culo a very effective shot on a standing animal that might move in any direction, and a culo also makes a nice loop on a moving animal because you can throw a big loop a long way, and the delivery doesn't need to be totally precise.

A culo loop can be thrown with the houlihan swing, a backhand swing, a forehand swing, a clockwise turnover or counterclockwise turnover swing. With all of these swings, you can still deliver a culo so that the tip of the loop comes in sometimes from underneath on the left side, from the right side, or sometimes from the right and left sides at the same time. In other words, there are a lot of variations when you're talking about culos!

"The houlihan and the forehand loops can be thrown in 360 degrees—from your horse's front, back, left or right sides."

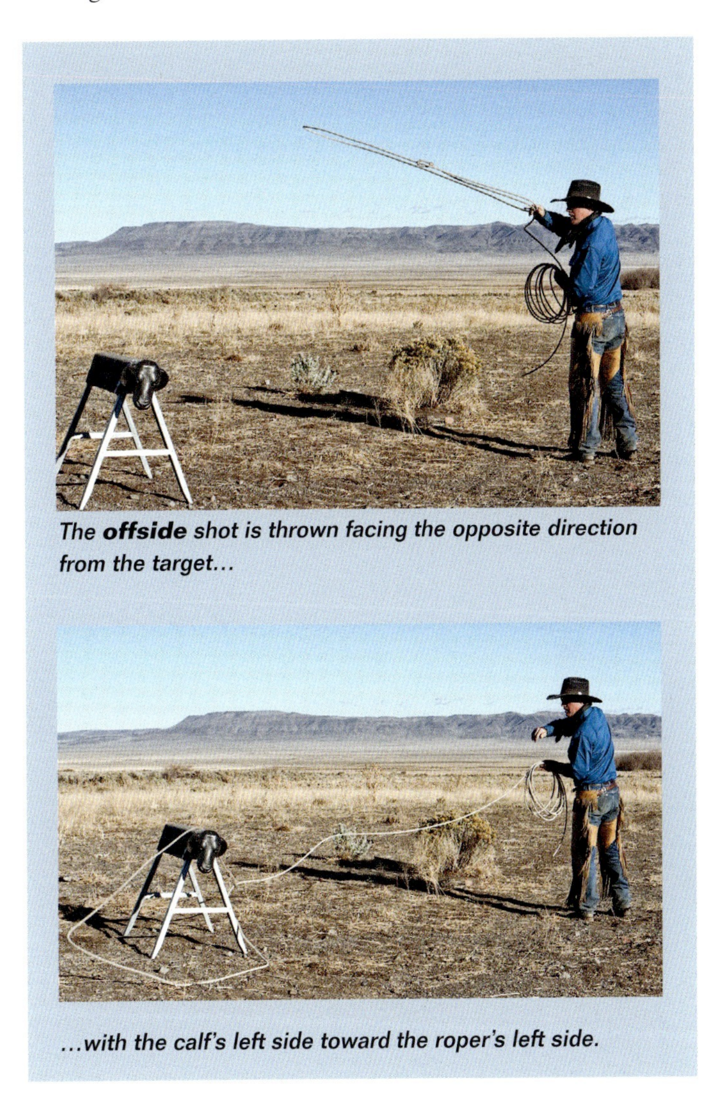

*The **offside** shot is thrown facing the opposite direction from the target...*

...with the calf's left side toward the roper's left side.

Backdoor: A favorite loop, the backdoor is any loop thrown toward a target that's back toward your horse's rump, or behind your stirrup. Only a few ropers deliver the loop this way, but the backdoor is very practical because, as discussed earlier, you can get deeper into the flight zone of an animal by approaching with your horse's tail instead of his head since this kind of approach is less threatening for the target animal.

When throwing a forehand backdoor loop off the right side, your elbow stays forward, and when your hand goes back, your palm should be up as you release the rope. Then you just turn your hand over and you're ready to dally.

You can deliver a backdoor shot using a forehand, backhand, or houlihan swing, and depending on your position, the loop can be either a head or heel shot. With a regular forehand delivery, your palm is down as your elbow comes back in the swing rotation. When your hand goes behind you, your palm rotates up, which creates a figure eight in your loop. As your hand comes forward, your palm comes down again, bringing the loop horizontal. Using the forehand swing, the backdoor loop should be delivered at the point where your loop normally goes through the figure eight.

If you want to prevent delivering a backdoor loop with a figure eight in the loop, leave your palm down as your hand comes back and throughout the delivery, even though you might deliver the loop behind you or off your right side as far as 3 or 4 o'clock.

In a backhand-swing delivery of a backdoor loop, your thumb ordinarily is up as you approach the release, so nothing changes. The same is true of a houlihan-swing delivery because your palm remains up with your thumb down all through the delivery.

*As this **forehand culo** leaves my hand ...*

... I rotate the loop counterclockwise...

... so the honda lands on the bottom and helps push the tip under the target.

Del Viento. Previously mentioned when discussing the clockwise turnover, a Del Viento shot is aimed above your target. After you deliver the loop, it rotates clockwise one time, falling down over the target while still rotating. This shot works well when trying to rope a calf that's hiding behind another cow, using it as a shield.

Reduce Stress, Increase Success

Any time you can ride quietly beside or among the animals you're trying to rope in the herd, this enables you to reduce stress on the cattle, and also gives you a more favorable shot. Remember that as long as you're behind the balance point, the animal moves forward. Always keep in mind where the flight zone and balance point are because you have better success if you can read and evaluate these while choosing your shot.

For example, if a cow wants to move away from you, stop approaching her directly and maintain your distance as you circle

Side-passing toward an animal can get a roper closer to a cow than he can get by coming with the horse's head first.

around her and get closer to make a throw. You create a lot of opportunity by staying outside the "flight zone" of the herd, each and every animal on the edge of the herd.

"The header actually has a good deal of responsibility when it comes to giving the heeler a successful shot."

Once the cattle start moving, that can create a ripple effect and make it difficult to set up that one shot on the edge.

*This **backhand culo** leaves my hand at about a 45-degree angle.*

The honda lands on top, and the top of the loop comes under from the left side.

When I'm in the herd, I don't like to follow a calf and rope him going away from me. Once the calf knows I'm following, self-preservation takes over and he looks for an escape. This can put the whole herd on alert. I prefer to stay out of my target's flight zone and rope him while he's standing still, or better yet, engage his curiosity and then rope him when he stops to look or takes a step toward me. Staying just outside the calf's flight zone often draws him away from the cows, and I can take advantage of his natural curiosity to set up a shot.

When you work range cattle that haven't been handled, you might find the calves' flight zones are bigger than the cows' flight zones. In this case, you can often use the cow as a shield to get closer to the calf and throw a loop over the cow to catch the calf. Or you can encourage the cow to move and fire a loop on the calf the second he sees you.

As you get close to a calf you're about to rope, this is where being able to totally control your horse's feet and make those half-steps we talked about in Chapter 3 come in handy. A full step might put you in the calf's flight zone so he turns and leaves, whereas a half-step might be all you need to get close enough to throw your loop.

Keep in mind that cattle are more intimidated by the front of your horse than by the horse's side or rump. You can sidestep or back a horse closer to a calf—without triggering his flight zone—than you can get by moving forward toward him. If you can rope from either side of your horse using a houlihan, backhand, offside, or backdoor throw, you find a lot more opportunities to get a closer shot.

When calves are gentle and used to being handled, you generally don't have any trouble getting close enough to catch them. It's when they want to hide and avoid you that you need to employ different methods.

Knowing how to work around the flight zone might require setting some coils in your hand with your loop. This can eliminate the drag of the coils coming out of your rein hand. Judge the distance to the calf and add enough coils to your right hand to cover the distance so that you don't need to drop coils from your left hand as you deliver the loop. Lay the coils flat in your roping hand so they don't tangle when you throw. As always, do the same with the coils in your rein hand so the first few coils can come out freely without getting fouled.

Tips for Headers

When you work in a large area outside and have a heeler who knows how to set up a shot, you shouldn't have a problem using a long rope. But when you work in closer quarters or don't have an experienced heeler, you want to keep your rope shorter. In any case, you always consider how much rope should used be when handling a calf in order to give the heeler the best shot.

Should the calf become scared once he's roped, it's usually best to keep the rope tight until the calf settles, then try to shorten your rope. Chasing a panicked calf to shorten

*For a **clockwise turnover**, as the top of the loop comes forward and up, pull it hard.*

This sends the loop above the target.

As seen in this photo, the loop rotates by itself...

... and as it starts down, I'm ready to gather my slack.

As it lands on the target, the loop continues to rotate until the honda is toward me.

*This **counterclockwise turnover** goes above the target...*

... rotates and drops down...

... to land with the honda toward me.

your rope can cause problems. Letting a calf settle down can make it easier for the heeler and the ground crew to perform their jobs. In most circumstances, if you get the calf caught and he starts pulling coils out of your hand, you probably need to dally and wait for the calf to settle.

When dragging a calf by the neck, going at a slow walk allows the calf to have at least three feet on the ground pulling against the rope. If you go fast enough that the calf must trot, this allows him to have only two feet on the ground at a time, so he can't pull back as much. The more feet the calf has on the ground, the more he can pull and the more he can choke. When he has more feet in the air, he can't pull back as much.

Some people might not like to see a calf handled this way with a head rope, and it can stress a calf when the rope tightens too much. However, done effectively, this method can actually be less stressful than running the calf around to heel it, separating cows and calves for hours and still catching the calf by the heels, or even running a calf through a chute.

Help for Heelers

No two situations are the same, and when you come in to heel a calf that is being dragged to the branding fire by the header, different strategies can improve your success.

The header actually has a good deal of responsibility when it comes to giving the heeler a successful shot. If the calf is gentle, the fire can be closer, and the heeler can get a good shot going straight to the fire. But if the calf is fresh and/or the heeler is inexperienced, it's easier to heel the calf when the header has more distance to go to reach the fire.

When the header's honda is under the calf's jaw, he can't choke much and the rope also lifts the animal's head so he can see the heeler coming. In this scenario, as a heeler you complicate your task when you come in behind the calf's balance point, as this drives him forward and makes it harder for the header to help you get a shot. But if you come into the flight zone ahead of the calf's balance point, where your stirrup is even with the calf's eye, slightly outside the radius of the rope, you're able to get closer. Then you can slow your horse and let the calf outtravel you and take the shot.

You want to be mindful of not getting inside the radius of the head rope; watch the angle of the header in relation to the calf. You need very little experience to fully understand the risk of being caught in the rope between the header and the calf.

When roping calves, it's important to understand that the flight zone of the calf is determined by its coherency. The coherency is determined by the oxygen level, and the oxygen level is determined by how much a rope restricts the calf's air intake. It's the responsibility of the ropers involved to carefully monitor this. To get a shot in a timely manner, a little stress at the right time can prevent a lot more stress later.

For example, when the header's honda is on top of the calf's neck, the animal might choke more readily, but the rope also pushes its head down, where it can't see behind as easily. In this case, you can come from straight behind to make your heel shot. The calf can be caught quickly, branded or doctored and then released, so the amount of stress the animal endures is shortened.

Obviously, you want to avoid a stressful scenario where the calf is choked too much. As heeler, your position on the balance

*The **backdoor** is thrown behind the roper, who has his palm up...*

... for a forehand catch. The backdoor also can be delivered backhand or houlihan.

point and how deep you are in the flight zone regulates how much the calf pulls against the head rope. The calf sits back more on the rope when you're in the flight zone and ahead of the balance point or are completely out of the flight zone even if you're behind the balance point.

"Used as a heel loop, the turn-over swing makes it easy to throw the loop on the opposite side of the calf. Then the loop comes back toward the roper, bringing the tip of the loop under the calf from the side opposite the rider."

The other extreme also should be avoided. In this case, the header catches the calf, but the heeler is in the calf's flight zone and behind the balance point. The calf will run up the header's rope to get air, which makes it hard for the heeler to get a shot, but more importantly, it also can lead to a situation where the calf gets exhausted from running so much. It takes only two or three such experiences, and the calf learns to lead and not set back on the rope, which makes it more difficult to get a consistent heel shot.

So be patient and wait for the right timing. You can get in a position that prepares the animal to give you a better shot by understanding how to set up the situation and wait for everything to come together. As you come in to take your shot, consider the fire's location and how your position regulates the calf's air intake.

A savvy roper is always conscious of the stock's best interest and monitors the calf's coherency and air intake. The ideal situation—and the least stressful for the calf—is one in which the animal's coherency level is the considering factor that allows the heeler to make a quick, efficient catch. This way, the whole ordeal is over for the calf in a matter of minutes, and then he can be turned loose to run back to his mama.

Practical Measures

Growing up and working in ranch country with few fences and fewer corrals, most of our stock was sorted and branded in the open without benefit of an enclosed area. Sometimes, we had the help of a fence corner or a large three-sided trap, but more often than not we roped and branded in an open rodear.

Because we couldn't rely on fencing, it was important to keep the cattle from becoming restless, and we learned to rope in ways that caused minimal disturbance to the herd. When this was done successfully, it took little, if any, effort to hold the herd. They were settled while we prepared the branding fire and, for the most part, they remained settled to the end. A great deal of responsibility for maintaining a quiet herd rested with the ropers who caught calves and brought them to the fire. The day's work could be frustrating or run smoothly, or be a combination, and this was largely due to the skills of those involved.

When we worked in an area with brush and rock, it usually was better to rope calves by the neck to take them to the fire. It's easier for a horse to pull a big calf bouncing on the end of a neck rope than it is for the horse to pull a calf on its side that has been roped by the heels. Once the calf got close to the fire, the heeler came in and caught the hind feet. Then the person on the ground doing the earmarking and castrating set the ropes to secure all four feet. It wasn't necessary for that person to return to the fire before performing his work. This method gave more ropers and horses a chance to develop and improve their skills than we'd have had if we used more men on the ground.

If we were shorthanded, the calves were small and the ground was clear, sometimes we opted to just heel calves and drag them to the fire. At that point, the front feet could be secured by a rope tied to a peg in the ground. No matter if we caught calves by the neck or dragged them by the heels, horses, not people on the ground, held them. Riders and horses alike were the better for it.

Valued Traditions

Like it or not, much of the way we do things today—including handling livestock—is all based on the clock and trying to get as much done as quickly as possible. When "hurry up and get done" is the main goal, a lot of value is lost.

Many of the methods of working cattle that we've covered here in these pages are a mystery to people, simply because they've never had the opportunity to try these methods and experience them.

I'm all for efficiency and completing the tasks that need to be done in a timely manner, but I don't think we have to sacrifice the cattle or the horses to do this. I also think we can continually improve our horsemanship and stockmanship. As long as we're around these animals on a regular basis, there is always something new to learn, as well as ways of doing things smarter, safer or easier than we did them before.

My hope is that horsemen and -women of all ages realize that efficient livestock handling can be accomplished by employing some valued traditions from earlier years. Learning to handle cattle effectively in open areas can help us develop the same feel we need to offer our horses. Forcing cattle in alleys and chutes involves the same brute force as getting a bigger bit to get a horse to do what we want. That might be effective and quick, but I personally enjoy learning the skills it takes to get things done without force.

Reading cattle can teach us a lot about reading horses, and handling cattle with horses gives our horses purpose. If I didn't have a purpose for my horses, I wouldn't be so interested in riding them; and if I'm not riding a horse, I have even less interest in working cattle. The purpose of working cattle on horseback is to improve the horse and make our jobs easier. It takes practice to develop these skills, and I hope some of this insight from my experiences makes other people's cattle-handling and riding experiences more pleasurable.

People might glamorize the old days, but there were also plenty of inconveniences and challenges back then. Rather than lament what is lost and long for the past, we can take lessons from those early buckaroos, cattlemen and ranchers, and incorporate those lessons into our daily work with livestock today. Some of those respected traditions still can fit seamlessly into our everyday work, enriching our lives and making routine tasks more rewarding.

As the calf approaches the ground crew, the heeler takes a shot that presents itself.

Glossary

backdoor loop — any loop thrown toward a target behind your stirrup, or toward your horse's rump.

balance point — the position from which you can influence an animal to change direction or stop. Balance point constantly changes, depending on your position to the animal.

culo — the Spanish term for "rump," which refers to any loop thrown at a cow's rump.

defense — when the horse is giving ground while holding the cow or decreasing pressure on the flight zone.

desensitization — the process of causing tolerance in the horse or cow by repeated exposure to various objects and/or situations.

dry work — exercises a stock horse does without using a cow.

flight zone — the area around an animal where, when something or someone approaches too closely, the animal feels the need to move and maintain a safe distance. The closer you are to the animal, the faster that animal moves. Likewise, the farther you move out of the flight zone, the more the animal slows or even stops.

honda (also, hondo) — the small reinforced loop at the end of a lariat or reata, through which the rope passes to form a loop.

"mother up" — when cows and calves, separated during work or when moving the herd, are allowed to find each other and "pair up," or get back together.

offense — when your horse moves toward a cow, pushing it away or increasing pressure on the cow's flight zone.

pair — a mother cow with her nursing calf.

pratha — a small group of cattle cut out and held apart from the main herd; also referred to as a "cut."

rango — the original Spanish term, still used in the Great Basin, for bringing in the horses, which has been Americanized to become "wrangle."

rating — refers to a horse "rating" a cow; a horse's natural tendency to track and follow a cow, which is similar to the way a foal follows its mother.

reata — a heavy rope made of braided or twisted rawhide.

rodear (also, rodea) — a herd of cattle held in the open and not contained by fencing.

PROFILE
CYNTHIA MCFARLAND

STEVE FLOETHE

Horse-crazy as long as she can remember, Cynthia McFarland was raised in Tucson, Arizona, where she spent countless hours riding the foothills and desert trails on the back of her trustworthy Quarter Horse gelding, Yuma.

A fulltime writer, Cynthia contributes regularly to a number of national equine publications. She is the author of eight books, and her writing has earned numerous awards, including a Steel Dust Award from the American Quarter Horse Association. Cynthia also freelances as a copywriter, writing text for Web sites, brochures and print ads, and has taught creative writing workshops for both children and adults.

When not writing, she enjoys riding the trails of north-central Florida on her good Paint horse, Ben. She and her four-legged family—Ben, Butler the donkey, beef cows and cats—live on a small farm in the horse country near Ocala.

Books Published by WESTERN HORSEMAN®

ARABIAN LEGENDS
by Marian K. Carpenter
280 pages and 319 photographs. Abu Farwa, *Aladdinn, *Ansata Ibn Halima, *Bask, Bay-Abi, Bay El Bey, Bint Sahara, Fadjur, Ferzon, Indraff, Khemosabi, *Morafic, *Muscat, *Naborr, *Padron, *Raffles, *Raseyn, *Sakr, Samtyr, *Sanacht, *Serafix, Skorage, *Witez II, Xenophonn.

BACKCOUNTRY BASICS
by Mike Kinsey with Jennifer Denison
212 pages, 180-plus color photographs. This problem-solving handbook is based on the award-winning Western Horseman magazine series for recreational riders, but anyone can benefit from this AQHA Professional Horseman's approach. Kinsey's step-by-step lesson plans develop a horse's confidence and willingness for a safer, more reliable and more responsive mount.

BACON & BEANS
by Stella Hughes
144 pages and 200-plus recipes for delicious Western chow.

BARREL RACING, *Completely Revised*
by Sharon Camarillo
128 pages, 158 photographs and 17 illustrations. Foundation horsemanship and barrel racing skills for horse and rider with additional tips on feeding, hauling and winning.

CHARMAYNE JAMES ON BARREL RACING
by Charmayne James with Cheryl Magoteaux
192 pages and 200-plus color photographs. Training techniques and philosophy from the most successful barrel racer in history. Vignettes that illustrate Charmayne's approach to identifying and correcting barrel-racing problems, as well as examples and experiences from her 20-plus years as a world-class competitor.

COWBOYS & BUCKAROOS
by Tim O'Byrne
176 pages and more than 250 color photograps. From an industry professional, trade secrets and the working lifestyle of these North American icons. The cowboy crew's four seasons of the cattle-industry year, cowboy and buckaroo lingo, and the Cowboy Code by which they live. How they start colts, handle cattle, make long circles in rough terrain and much, much more, including excerpts from the author's journal.

FIRST HORSE
by Fran Devereux Smith
176 pages, 160 black-and-white photos and numerous illustrations. Step-by-step information for the first-time horse owner and/or novice rider.

HELPFUL HINTS FOR HORSEMEN
128 pages and 325 photographs and illustrations. *WH* readers' and editors' tips on every facet of life with horses. Solutions to common problems horse owners share. Chapter titles: Equine Health Care; Saddles; Bits and Bridles; Gear; Knots; Trailers/Hauling Horses; Trail Riding/ Backcountry Camping; Barn Equipment; Watering Systems; Pasture, Corral and Arena Equipment; Fencing and Gates; Odds and Ends.

IMPRINT TRAINING
by Robert M. Miller, D.V.M.
144 pages and 250 photographs. How to "program" newborn foals.

LEGENDARY RANCHES
By Holly Endersby, Guy de Galard, Kathy McRaine and Tim O'Byrne
240 pages and 240 color photos. Explores the cowboys, horses, history and traditions of contemporary North American ranches. Adams, Babbitt, Bell, Crago, CS, Dragging Y, Four Sixes, Gang, Haythorn, O RO, Pitchfork, Stuart and Waggoner.

LEGENDS 1
by Diane C. Simmons with Pat Close
168 pages and 214 photographs. Barbra B, Bert, Chicaro Bill, Cowboy P-12, Depth Charge (TB), Doc Bar, Go Man Go, Hard Twist, Hollywood Gold, Joe Hancock, Joe Reed P-3, Joe Reed II, King P-234, King Fritz, Leo, Peppy, Plaudit, Poco Bueno, Poco Tivio, Queenie, Quick M Silver, Shue Fly, Star Duster, Three Bars (TB), Top Deck (TB) and Wimpy P-1.

LEGENDS 2
Various Authors
192 pages and 224 photographs. Clabber, Driftwood, Easy Jet, Grey Badger II, Jessie James, Jet Deck, Joe Bailey P-4 (Gonzales), Joe Bailey (Weatherford), King's Pistol, Lena's Bar, Lightning Bar, Lucky Blanton, Midnight, Midnight Jr, Moon Deck, My Texas Dandy, Oklahoma Star, Oklahoma Star Jr., Peter McCue, Rocket Bar (TB), Skipper W, Sugar Bars and Traveler.

LEGENDS 3
Various Authors
208 pages and 196 photographs. Flying Bob, Hollywood Jac 86, Jackstraw (TB), Maddon's Bright Eyes, Mr Gun Smoke, Old Sorrel, Piggin String (TB), Poco Dell, Poco Lena, Poco Pine, Question Mark, Quo Vadis, Royal King, Showdown, Steel Dust and Two Eyed Jack.

LEGENDS 4
Various Authors
216 pages and 216 photographs. Blondy's Dude, Dash For Cash, Diamonds Sparkle, Doc O'Lena, Ed Echols, Fillinic, Harlan, Impressive, Lady Bug's Moon, Miss Bank, Miss Princess/Woven Web (TB), Rebel Cause, Tonto Bars Hank, Vandy, Zan Parr Bar, Zantanon, Zippo Pine Bar.

LEGENDS 5
Various Authors
248 pages, approximately 300 photographs. Bartender, Bill Cody, Chicado V, Chubby, Custus Rastus (TB), Hank H, Jackie Bee, Jaguar, Joe Cody, Joe Moore, Leo San, Little Joe, Monita, Mr Bar None, Pat Star Jr., Pretty Buck, Skipa Star, and Topsail Cody.

LEGENDS 6
Various Authors
236 pages, approximately 270 photographs. Billietta, Caseys Charm, Colonel Freckles, Conclusive, Coy's Bonanza, Croton Oil, Doc Quixote, Doc's Prescription, Dynamic Deluxe, Flit Bar, Freckles Playboy, Great Pine, Jewels Leo Bars, Major Bonanza, Mr San Peppy, Okie Leo, Paul A, Peppy San, Speedy Glow and The Invester.

LEGENDS 7
Various Authors
260 pages and 300-plus photos. Big Step, Boston Mac, Commander King, Cutter Bill, Doc's Dee Bar, Doc's Oak, Gay Bar King, Hollywood Dun It, Jazabell Quixote, Mr Conclusion, Otoe, Peppy San Badger, Quincy Dan, Rey Jay, Rugged Lark, Skip A Barb, Sonny Dee Bar, Te N' Te, Teresa Tivio and War Leo.